$\mathscr{R}$ Gas constant

$\mathscr{S}$ Entropy

S Solubility

S^0 Solute adsorption energy

$S_{B/A}$ Separation factor (for B)

T Temperature:

> T_a = ambient temperature
>
> T_c = column temperature
>
> T_f = final temperature

T' Significant temperature (in PTGC)

T_r Throughput

U Nonequilibrium concentration

V Volume:

> V_A = molar volume of solute A
>
> V_{extr} = volume of extractant
>
> V_{raff} = volume of raffinate
>
> V_a = adsorbent surface volume
>
> V_M = volume of the mobile phase
>
> V_S = volume of the stationary phase
>
> V_G = volume of the gas phase
>
> V_L = volume of the liquid phase
>
> V_T = total volume
>
> V_{ss} = volume of the solid support
>
> V_R = retention (elution) volume

> V_{R^0} = corrected retention volume
>
> V_N = net retention volume
>
> V_g = specific retention volume
>
> $\bar{V}$ = molal volume of of a liquid

W Weight; for example W_A

X Mole fraction; for example X_A

Y Segregation factor

Z Miscellaneous constant

a Activity; for example a_A

b Number of phases (phase rule)

c Number of components (phase rule)

d Distance between two peak maxima

d_f Film thickness

d_p Particle diameter

f Fugacity

f_{r,n_t} Fraction of solute in funnel r after transfer n_t

g Number of degrees of freedom (phase rule)

h Reduced H

j Pressure correction factor

k Proportionality constant (general) or kinetic rate constant or permeability coefficient

k_B Boltzmann constant

k' Partition ratio

l Length of a random walk step or fractional part of L

m Weight of sorbent

continued on back

SEPARATION METHODS
IN CHEMICAL ANALYSIS

SEPARATION METHODS
IN CHEMICAL ANALYSIS

James M. Miller

Drew University
Madison, New Jersey

A WILEY-INTERSCIENCE PUBLICATION

JOHN WILEY & SONS

New York · London · Sydney · Toronto

Framingham State College
Framingham, Massachusetts

Copyright © 1975, by John Wiley & Sons, Inc.

Library of Congress Cataloging in Publication Data:

Miller, James M.
  Separation methods in chemical analysis.

  "A Wiley-Interscience publication."
  1. Separation (Technology)   2. Chemistry, Analytic.
I. Title.

QD63.S4M54   544'.1   74-13781
ISBN 0-471-60490-9

Printed in the United States of America

10 9 8 7 6 5 4 3 2 1

PREFACE

Separation methods have become an important part of analytical chemistry and should be included in the undergraduate chemistry curriculum. This book was developed for the course "Separations and Analysis," which has been offered at Drew University since 1970. For schools without a separate course in separations, this text should be useful in analytical and instrumental courses at the undergraduate and dual levels. Industrial chemists should find it helpful as a summary of recent developments in separation science.

The period just prior to 1970 saw a tremendous growth in separation methods, including extensive exploitation of gas chromatography and the beginning of modern liquid chromatography. Unfortunately, each separation method was developing its own theories and nomenclature without much regard for the other methods, thus making comparisons between methods difficult. Attention was focused on this problem at an NSF-sponsored short course entitled "The Theory of Chromatography—A Unified Approach," held at Drew in the summer of 1968. There it became apparent that a unified approach was both necessary and possible, not only for chromatographic separations but for all separation methods.

This book is intended to serve as a unifying outline of the major methods of separation used in the analytical laboratory, to present the basic theory of separation science, and to provide a framework into which new and additional methods can be fitted. It should facilitate comparisons between methods and aid in the selection of the best method for a given analysis, within the limitations imposed by its brevity.

It is assumed that the student has had some introduction to elementary thermodynamics, equilibrium principles, and the formation of coordination complexes, so these subjects are treated very briefly. Furthermore, the student has probably had some prior laboratory experience with distillation, batch extraction, chromatography (probably thin layer or paper chromatography), precipitation, and absorption spectroscopy, so it is assumed that they can be discussed before extensive descriptions are given.

The book contains over 300 references, most of them to the primary literature, and students should be encouraged to consult them. By and large, if a subject is not discussed in the text, a reference is given to provide the starting place for a search for further information.

I have tried very hard to be consistent in the use of terms and symbols, although the choice of a set of symbols was very difficult due to the lack of uniformity between fields. Most symbols were taken from the field of gas chromatography because of the attention workers in that field have paid to uniformity. Students should be cautioned, however, that in reading various texts and papers the symbols vary widely and can be confusing.

While this text was in preparation a similar one appeared (Karger, Snyder, and Horvath, *An Introduction to Separation Science*, Wiley, New York, 1973). It is more advanced but has the same goals of unification and hence is similar in many respects. Because of the timing, it was not possible to cross reference this book extensively with Karger's but in general it is recommended as a more advanced resource book.

I want to express my appreciation to a large number of people, beginning with the many scientists on whose work I have drawn in the preparation of this text. This is especially true for those lecturers and participants at the 1968 short course mentioned earlier. My students who suffered through earlier versions of this text deserve thanks for their perseverence and patience as well as their helpful criticisms. I also want to thank reviewers, other friends (individuals and corporations) who have encouraged and helped me, the many typists, and Drew University. Special thanks are due my family and Leslie Ettre.

JAMES M. MILLER

Madison, New Jersey
August 1974

CONTENTS

Instrumentation
Retention Parameters and Qualitative Analysis
Classification of Liquid Phases
Temperature Effects
Optimization
Quantitative Analysis
Miscellaneous Topics
Evaluation

Classical LC
High-Pressure LC Apparatus
Theory
Miscellaneous Topics

Liquid–Solid Chromatography
Liquid–Liquid Chromatography
Ion-Exchange Chromatography
Gel-Permeation Chromatography

Theory
Apparatus and Techniques
Comparison of Thin Layer and Paper Chromatography
Comparison of LC Separations in Columns and on Plane Surfaces
Summary

Theory
Apparatus and Techniques
Summary and Evaluation

Theory
Apparatus and Procedures
Summary

INTRODUCTION

One of the most rapidly developing areas of analytical chemistry is separation methods, mainly due to the advent of gas chromatography (GC), a method of high efficiency and relatively low cost. The development of GC, beginning in the 1950s, has made possible many new analyses and permitted the improvement of others. The result has been an increased interest in, and comprehension of, many separation methods. A few examples of recent triumphs of GC will serve to illustrate the advances being made in analytical chemical separations.

The petroleum industry was the first to make extensive use of GC, and it soon became the most important analytical method in the industry. Only 11 years after the first GC paper, the nearly complete separation and identification of all 39 hydrocarbons in the 28 to 114°C boiling fraction was reported (*1*), and by 1968 an average gasoline could be shown to contain about 240 compounds, of which 180 had been identified (*2*). Fractional distillation, the method previously used for petroleum separations, was quickly replaced by GC for many applications. In fact, GC has even been used to simulate distillation (*3*).

The impact of GC on other fields has been equally impressive. Studies of environmental pollution are often carried out by GC. They range from the parts-per-billion analysis of sulfur compounds in air (*4*) to the analysis of polychlorinated biphenyl (PCB) pesticide residues at the nanogram level (*5*) and the identification of hydrocarbon pollutants on seas and beaches (*6*). Some interesting studies on humans include the analysis of effluents picked up by a stream of air passing over the human body; this yields chromatographic "profiles" that are characteristic enough to distinguish between male and female (*7*). Other profiles have been obtained from volatiles extracted from urine. About 300 compounds have been detected, and 40 of them have been identified (*8*). It is hoped that both types of profile will aid in early diagnosis of disease in the near future.

Another application is the analysis of drugs in urine. At the 1972 Olympics in Munich more than 2500 randomly selected athletes were screened to make sure that they were not taking drugs to increase their prowess in the games. About 30 banned drugs were sought, including ephedrine, but not caffeine.

The former drug, a common ingredient in antihistamine formulations, was being taken by an American swimmer on his doctor's orders. When the drug was detected in his urine, the young man was disqualified and stripped of the medal he had won; this caused considerable controversy (9). A sophisticated GC method was the one being used because it was fast and sensitive.

The theory of chromatography has also developed rapidly during the last 20 years, providing the fundamentals needed to stimulate advances in the practical applications. Much of the theory is applicable to other chromatographic techniques and indeed to other nonchromatographic separation methods. Consequently the field of separations is no longer just an art but rather a science. New monographs, treatises, and journals have been published to disseminate this knowledge, which is becoming known as *separation science*. Especially in this formative period it is important that separations be treated on a unified basis with common symbols to emphasize the similarities and differences between the methods. Only a unified treatment facilitates the transfer of information from one field to another.

The purpose of this book is to provide a unified study of the most common separation techniques used in the analytical laboratory. It emphasizes the organization of material and presents a framework into which separation techniques can be fitted. The next chapter presents classification schemes and definitions needed to begin the study. As the specific techniques are presented in the chapters that follow, Chapter 2 should take on more meaning until, at the end, it can be reconsidered as a good summary.

Chapters 3 through 6 present fundamentals that are common to all separations. Chapter 7 is the first of nine chapters devoted to specific separation methods. Chapters 10 through 17 are also on specific methods, but they are preceded by Chapters 8 and 9, which discuss some additional fundamentals they have in common. Finally, in Chapter 18 the methods are compared and suggestions are made for selecting the best one for a given problem. In approximate decreasing order of importance in the text, the separation methods covered are the following:

1. Gas chromatography
2. Liquid chromatography in columns
3. Liquid chromatography on plane surfaces
4. Distillation
5. Liquid–liquid extraction
6. Zone electrophoresis
7. Dialysis
8. Precipitation
9. Formation of inclusion compounds
10. Masking

THE RELATIONSHIP BETWEEN SEPARATION AND ANALYSIS

In some analyses, the measurement, whether qualitative or quantitative, can be made directly on the sample. More often, however, a separation step is necessary to permit the measurement to be made in a medium free of any interferences. When such is the case, the separation step is often the most difficult one in the analysis.

The relationship of the separation step to the entire process is apparent from the following list of the steps in a quantitative analysis:

1. Selecting and preparing the sample
2. Measuring the sample
3. Dissolving the sample
4. Preliminary treatment, such as adjusting the pH
5. Separating the desired constituent
6. Measuring the desired constituent
7. Analyzing the data and reporting

In some chromatographic techniques, especially GC, the measurement step is closely associated with the separation step, and the chromatographic instrument includes both functions. For this reason, Chapter 12 contains a section on quantitative analysis, but the measurement step is not discussed in conjunction with any of the other separation methods. For the most part, it will be more instructive to distinguish clearly between separation and measurement.

Finally, this text is aimed at laboratory-scale separations, emphasizing the analytical applications. It does not cover process or production-scale operations, which would require a broader treatment.

REFERENCES

1. R. D. Schwartz and D. J. Brasseaux, *Anal. Chem.* **35,** 1374 (1963).
2. W. N. Sanders and J. B. Maynard, *Anal. Chem.* **40,** 527 (1968).
3. L. E. Green, L. J. Schmauch, and J. C. Worman, *Anal. Chem.* **36,** 1512 (1964).
4. F. Bruner, A. Liberti, M. Possanzini, and I. Allegrini, *Anal. Chem.* **44,** 2070 (1972).
5. E. J. Bonelli, *Anal. Chem.* **44,** 603 (1972).
6. E. R. Adelard, L. F. Creaser, and P. H. D. Matthews, *Anal. Chem.* **44,** 64 (1972).
7. *Chem. Eng. News*, November 6, 1972, p. 60; December 18, 1972, p. 40.
8. A. Zlatkis, paper presented at the 164th ACS Meeting, New York, August 29, 1972.
9. *Anal. Chem.* **44,** [12], 77A (1972); *New York Times*, November 10, 1972, p. 27.

DEFINITIONS AND CLASSIFICATIONS 2

As a separation method develops, insufficient attention is usually paid to the formulation of definitions and the assignment of symbols. Consequently each method tends to have its own terms and symbols, which have little meaning for other methods. Even worse is the situation where the *same* term has *different* meanings in different techniques.

This confusion must be reduced to a minimum if a unified presentation is to be successful. Choosing terms and symbols is very difficult, however, since one is not right and the other wrong, and there is considerable room for personal opinion and preference.

In this volume I use, with a few exceptions, the symbols and definitions that originated in GC because they seem to have gained the widest general acceptance in the United States. Other choices were made on the basis of what seemed logical in the light of my objectives. Many will disagree with the choices made since some differ from common practice. For these reasons the definitions and symbols should be carefully considered. Some meanings may be different from those with which you are familiar. Similar caution should be exercised in reading other monographs and papers since this may cause some difficulty but will make you aware of the problems of semantics in science.

BASIC DEFINITIONS

Separation

Since "separation" is a familiar word, it is unlikely that any confusion will arise from its use in reference to chemical analysis. Nevertheless, there is some merit in considering a precise, general definition such as that suggested by Rony (1): "Separation is the hypothetical condition where there is complete isolation, by m separate macroscopic regions, of each of the m chemical components which comprise a mixture. In other words, the goal of any separation process is to isolate the m chemical components, in their pure forms, into m separate vessels, such as glass vials or polyethylene bottles." The adjective "hypothetical" is used for two reasons. In the first place, it is theoretically impossible to accomplish the *complete* separation of the components of a mixture. This will be made clear in later discussions. In the second place, it is often the case that the separated compounds are not actually isolated, but rather a record (e.g., on chart paper) contains the information that indicates how well the components were separated.

Consider constituents A and B present together in a homogeneous phase. In order to separate A and B, a second phase will have to be added or formed, so that in the end A is contained in one phase and B in the other. Thus a separation system is usually composed of two phases. (In more complicated systems additional phases may be present, but that is not important at this point.) Usually the distribution of the components between the two phases is allowed to come to equilibrium and the theory can be presented in terms of equilibrium constants. After equilibrium is reached, the phases are mechanically separated and the process is complete.

For example, in liquid–liquid extraction (LLE) the two phases are immiscible liquids, and the sample components partition or distribute themselves between the two phases, depending primarily on their relative activities (solubilities) in the two phases. Then the phases are separated using a separatory funnel or treated filter paper so that the components can be recovered and/or measured. This final, mechanical separation step is usually simple and obvious for all separation processes and does not need extensive discussion.

The Phases

For simplicity the two phases will be numbered 1 and 2. However, in some techniques they have special names, and these are listed in Table 1 to emphasize their similarities. For example, the stationary phase in chromatography is similar in function to the raffinate in extraction and the retentate in dialysis.

Table 1. Names of Phases in Some Separation Methods

Technique	Phase 1	Phase 2
Chromatography	Stationary	Mobile
Liquid–liquid extraction	Raffinate	Extractant
Dialysis	Retentate	Diffusate

Other Terms

Although many authors distinguish between them, the terms "partition" and "distribution" are used synonymously in this volume to describe the general process by which a system comes to equilibrium. The two subclassifications are (1) *ad*sorption, which describes distribution processes occurring at

ABsorption   ADsorption

Figure 1.   The difference between absorption and adsorption.

interfaces; and (2) *ab*sorption, which describes distribution processes occurring in the bulk phases.* A comical illustration of this distinction is shown in Figure 1.

The Equilibrium Constant

If a separation system is allowed to come to equilibrium, the ratio of the concentrations (activities) of each component (solute) in the two phases becomes constant and can be expressed as an equilibrium constant, called a partition or distribution coefficient, K_D. Using A and B to represent solutes (sample components) and 1 and 2 to represent phases, K_D is

$$(K_D)_A = \frac{(C_A)_1}{(C_A)_2} = \frac{\text{concentration of A in phase 1}}{\text{concentration of A in phase 2}} \tag{1}$$

* Those familiar with GC will recognize that these are *not* the definitions commonly used in that field. In GC, adsorption has the same meaning, but partition (not absorption) is used to describe bulk-phase distribution in gas–liquid chromatography.

for component A and

$$(K_D)_B = \frac{(C_B)_1}{(C_B)_2} \tag{2}$$

for component B.†

The symbol C is used to represent the "analytical" concentration; that is, it includes all forms (ionic, molecular, etc.) in which A occurs in each of the phases. Consequently, the value of K_D can change with concentration as the relative amounts of the various forms of the solute change.

On the other hand, a partition coefficient can be written for only one of the forms of the solute—for example, the molecular form. The symbol K_p is used to denote it:

$$(K_p)_A = \frac{[A]_1}{[A]_2} \tag{3}$$

In this case brackets indicate molar concentrations of the single species. To simplify subsequent discussions of partition coefficients, it will be assumed that the solute occurs in only one form, and K_p can be used.

For the partition coefficient to be a true constant, it is necessary to define it in terms of activity instead of concentration:

$$(K_p)_A = \frac{(a_A)_1}{(a_A)_2} = \frac{[A]_1(\gamma_A)_1}{[A]_2(\gamma_A)_2} \tag{4}$$

where γ is the activity coefficient. Equation 4 is equal to equation 3 if the activity coefficients are unity. This assumption is made unless otherwise noted in order to simplify the discussion. Activity coefficients are discussed in the next chapter.

Returning to the basic definitions, the partition coefficient can be divided into two terms as follows:

$$(K_p)_A = \frac{[A]_1}{[A]_2} = \frac{(W_A)_1/MW_A}{V_1} \div \frac{(W_A)_2/MW_A}{V_2} \tag{5}$$

where W_A is the weight of A (in grams). MW_A is the molecular weight of solute A, and V is the volume of the respective phase. Since the molecular

† According to the definitions of phases in Table 1, the partition coefficient for liquid–liquid extraction is

$$K_D = \frac{(C)_{raff}}{(C)_{extr}}$$

It should be noted that this is the reciprocal of the coefficient as normally used (extractant over raffinate), where the raffinate is taken as the aqueous phase and the extractant as the organic (nonaqueous) phase.

weight of A cancels,

$$(K_p)_A = \frac{(W_A)_1}{(W_A)_2} \frac{V_2}{V_1}$$ (6)

The term $(W_A)_1/(W_A)_2$ is the ratio of the total amount (weight) of A in one phase to that in the second phase. It is called the partition ratio or capacity factor and is given the symbol k'. It also represents the ratio of mole fractions:

$$k' = \frac{(W_A)_1}{(W_A)_2} = \frac{(X_A)_1}{(X_A)_2}$$ (7)

The ratio of the two volumes of phases is called the phase ratio and is given the symbol β:

$$\beta = \frac{V_2}{V_1}$$ (8)

Thus

$$K_p = k'\beta$$ (9)

Equation 9 states one of the important, fundamental relationships for those separation methods that can be described by equilibrium theory.

DESCRIPTION OF MAJOR SEPARATION METHODS

The separation methods that receive most attention in this volume are listed in Chapter 1. Brief descriptions of the most important ones are included here to serve as a necessary survey for the material that follows. A few fundamental equations are also presented.

Liquid–Liquid Extraction

Liquid–liquid extraction (LLE) is a simple and popular technique. Two immiscible (or very slightly miscible) liquids are used as the two phases. In the simplest case, the sample is dissolved in one of the liquids (the raffinate), which is contacted with the other liquid (the extractant) in a separatory funnel that is shaken gently to increase the contact between phases. The components of the sample distribute themselves between the two phases and come to equilibrium. Ideally, one component stays primarily in the raffinate, and the other goes into the extractant. Thus a separation is effected. Other methods of extraction, including the more complex multistep countercurrent process, are described in Chapter 10.

It is often convenient to express the extent of the extraction for a single solute as the fraction extracted, Q. Thus the fraction unextracted would be $(1 - Q)$. The ratio of the fraction unextracted to the fraction extracted $(1 - Q)/Q$ is the same as the ratio of the amount of solute in the raffinate (unextracted) to the amount in the extractant. This is the partition ratio k':

$$\frac{1 - Q}{Q} = \frac{(W)_1}{(W)_2} = \frac{W_{\text{raff}}}{W_{\text{extr}}} = k' \tag{10}$$

Thus from equation 9,

$$K_p = \frac{1 - Q}{Q} \frac{V_2}{V_1} \tag{11}$$

Rearranging equation 11 gives

$$Q = \frac{V_2}{V_2 + K_p V_1} = \frac{1}{1 + K_p/\beta} = \frac{1}{1 + k'} \tag{12}$$

and, for the fraction unextracted,

$$(1 - Q) = 1 - \frac{V_2}{V_2 + K_p V_1} = \frac{V_2 + K_p V_1 - V_2}{V_2 + K_p V_1} = \frac{K_p V_1}{V_2 + K_p V_1} = \frac{k'}{k' + 1} \tag{13}$$

If it is assumed that the volumes of the two phases are equal $(\beta = 1)$, equations 12 and 13 respectively become

$$Q = \frac{1}{K_p + 1} \tag{14}$$

and

$$(1 - Q) = \frac{K_p}{K_p + 1} \tag{15}$$

These equations are simple to work with and are used in most of the theoretical discussions.

Chromatography

Although the chromatographic system may comprise more than two phases, two are of primary interest in describing the technique. One phase is an active sorbent that remains stationary, and the other phase is mobile and percolates through the stationary phase. The sample is placed at one end of the stationary sorbent bed (which may be packed in a column) and washed continually with fresh mobile phase. As the sample is washed down the bed by the mobile phase, the various components of the sample are retarded in proportion to their interaction with (sorption on) the sorbent bed—that is, the sample partitions between the two phases. At any given time, a particular solute molecule is either in the mobile phase, moving along at its velocity, or in the stationary phase and not moving at all. The sorption–desorption process

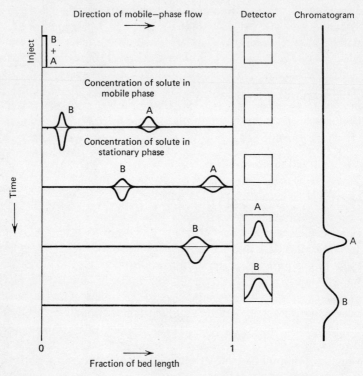

Figure 2. Schematic representation of the chromatographic process.

occurs many times as the molecule moves through the bed, and the time required to do so depends primarily on the length of time it is sorbed and held immobile. A separation is effected as the various components emerge from the bed at different times. The process is depicted in Figure 2.

This type of chromatographic development is called elution chromatography since the sample is continuously washed or eluted down the column by the mobile phase. By contrast, in displacement chromatography the sample is displaced down the column by the mobile phase (and by other components in the sample). It is a less popular technique, and all subsequent discussion refers to the elution process unless otherwise noted.

This type of chromatography could also be called zonal chromatography because the sample is applied to the bed all at once in a narrow zone (also referred to as a spike). By contrast, it is possible to apply the sample continuously during the run. This type is usually called frontal analysis chromatography, and, like displacement, it is less popular.

The various chromatographic techniques are often referred to by the states of the two phases, naming the mobile phase first. Thus we have gas–liquid (GLC), gas–solid (GSC), liquid–liquid (LLC), and liquid–solid (LSC) chromatography. Another distinction is based on the shape of the stationary bed, which can be contained in a column (column chromatography) or coated on a flat surface (plane chromatography). The two common types of plane chromatography are paper and thin-layer chromatography (TLC).

Although chromatography is a dynamic process, it has been treated by equilibrium theory, and the equilibrium partition coefficient appears in the basic chromatographic equation:

$$V_R^0 = V_M^0 + K_p V_S \tag{16}$$

where V_R^0 is the volume of mobile phase required to elute the sample from the bed and is called retention volume or elution volume.* V_M^0 and V_S are the volumes of the mobile and stationary phases, respectively, and K_p is the partition coefficient previously defined. Equation 16 states that the total volume of mobile phase that flows while a given solute passes through the bed is the sum of two parts: V_M^0 represents the "dead" volume in the bed through which every solute must pass, and $(K_p V_S)$ represents the mobile phase that flows while the solute is immobile in the stationary phase. Obviously the latter depends on the affinity of the solute for the stationary phase (expressed by K_p) and the amount of stationary phase (V_S). The derivation of this equation is found in Chapter 11.

Although equation 16 is one of the fundamental equations of chromatography, it is more common for students to be introduced to chromatographic theory via another parameter, the retardation factor, or R_F value. It is used to express retention data in paper chromatography and thin-layer chromatography and is defined as

$$R_F = \frac{\text{distance a solute migrates through the bed}}{\text{distance the solvent front migrates through the bed}} \tag{17}$$

The R_F values are always equal to, or less than, unity because a solute cannot migrate farther than the solvent front migrates. The more a sample is retarded

---

* The zero superscript in the terms V_R^0 and V_M^0 has a specific meaning and name in GC (see Chapter 12). In liquid chromatography it is not necessary, but, for the unification of these two forms of chromatography, the superscript will be used in all general chromatographic equations where it would be appropriate for GC.

by the bed, the smaller will be its R_F value (and the larger would be its retention volume, V_R^0). The R_F, rather than V_R^0, values are used because it is easier to measure *distances* on a piece of paper or a TLC plate than it is to measure the *volume* of solvent. It is obvious, however, that there must be a relationship between R_F and V_R^0.

To arrive at this relationship we need to define another type of retardation factor, similar to, but not exactly the same as, R_F. It is called the retention ratio R_R, defined as

$$R_R = \frac{\text{average velocity of the solute through the bed}}{\text{average velocity of the solvent through the bed}} = \frac{\bar{v}}{\bar{u}} \qquad (18)$$

It differs from R_F in two respects: first, velocities rather than distances are measured, and consequently R_R is preferred over R_F for theoretical discussions; second, the *denominator* relates to the bulk solvent velocity rather than the solvent *front* as is the case with R_F.*

The relationship between R_R and V_R^0 is inherent in the definition of R_R itself. The velocity of the solute, $\bar{v}$, is given by

$$\bar{v} = \frac{L}{t_R^0} \qquad (19)$$

where L is the length of the bed and t_R^0 is the solute retention or elution time. Similarly, for the solvent velocity, $\bar{u}$,

$$\bar{u} = \frac{L}{t_M^0} \qquad (20)$$

where t_M^0 is the retention time for a nonsorbed sample and represents the time required to pass through the bed without stopping. Combining equations 18, 19, and 20, we obtain

$$R_R = \frac{\bar{v}}{\bar{u}} = \frac{L/t_R^0}{L/t_M^0} = \frac{t_M^0}{t_R^0} \qquad (21)$$

This equation defines the retention ratio as the fraction of time a solute spends in the mobile phase—a valid concept of its own.

Assuming a constant flow rate F, equation 21 can be converted to the equivalent volumes since $V = Ft$:

$$R_R = \frac{V_M^0}{V_R^0} = \frac{V_M^0}{V_M^0 + K_p V_S} \qquad (22)$$

* Giddings (2) has pointed out that an R_F value for a given system will always be smaller than the corresponding R_R value since the solvent *front* moves faster than the bulk solvent and the denominator in the R_F definition will be larger than for R_R. The reason for the velocity differences is that the solvent front moves on a dry bed, whereas the bulk solvent moves on the previously wetted bed.

This is the relationship sought for. Furthermore,

$$R_R = \frac{1}{1 + K_p/\beta} = \frac{1}{1 + k'} \quad (23)$$

and

$$(1 - R_R) = \frac{K_p V_S}{V_R^0} = \frac{k'}{1 + k'} \quad (24)$$

For the special case when the phase volumes are equal, $\beta = 1$, equations 22 and 23 respectively reduce to

$$R_R = \frac{1}{1 + K_p} \quad (25)$$

and

$$(1 - R_R) = \frac{K_p}{1 + K_p} \quad (26)$$

Note the similarities between these last four equations and equations 12 to 15 for LLE. Clearly the concept of "fraction extracted" (Q) in LLE is related to that of "retention ratio" (R_R) in chromatography. Also, it should be remembered that these equations apply to the "average" molecule for a given solute. The effect of the separation process on the large number of molecules actually present in a sample is covered in Chapter 8.

Distillation

In distillation one begins with a single liquid phase that contains some volatile components. When it is heated, a second phase is formed, which is a vapor richer in the volatile components. When the vapor phase is condensed, the resulting liquid is richer in the volatile components and a separation has been effected based on volatility. The condensed liquid is called the distillate and the remaining undistilled (less volatile) liquid is called the reflux. A diagram of the apparatus is shown in Figure 3.

To make the process more efficient, the area of contact between the liquid and the vapor can be increased by using fractionating columns like those shown in Figure 4. Some of the vapor condenses and runs down the column while contacting the vapors moving up the column. Equilibration takes place, resulting in distillate that is rich in the volatile component. This type of distillation is called fractional distillation, and it is the type of interest in the laboratory.

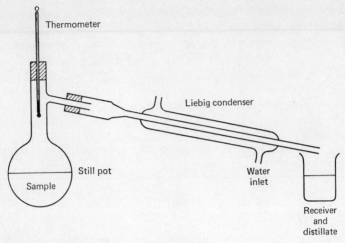

Figure 3. Typical apparatus for distillation.

Electrophoresis

As the name implies, the basis of electrophoresis is the movement of charged species in a potential gradient. As in chromatography, there are two funda-mental types: frontal (a moving boundary) and zonal. In practice, the frontal technique is usually carried out on an unsupported solution (free) while the zonal is carried out on a support such as paper. Hence these two sets of terms tend to be used interchangeably and one often finds the two types of electro-phoresis categorized as "free" and "zonal." It is *zonal electrophoresis on a support* that is of primary interest in the analytical laboratory.

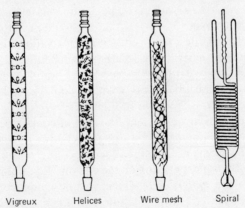

Figure 4. Typical fractionating columns for distillation.

Since the support used in zone electrophoresis is similar to the bed used in plane chromatography, this technique is also known as electrochromatography. Strictly speaking, zone electrophoresis refers to a differential migration of solutes from an initial zone under the influence of an electrical field, and interaction between the sample and the support is not intended; thus electrophoresis is not a chromatographic process at all. In actual operation, however, it is difficult to prevent some interaction between the sample and the support, so that the net result is a combination of electrophoresis and electrochromatography. Since in practice one cannot separate the effects, it is common to use the term "zone electrophoresis" to describe the technique rather than process and to acknowledge a certain amount of chromatographic effect.

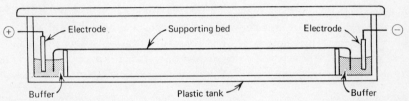

Figure 5. Simple apparatus for zone electrophoresis.

A typical apparatus for zone electrophoresis is shown in Figure 5. The paper support is wetted with the solvent and dipped into reservoirs of it. The solvent must be an electrolyte. The sample is applied to the support, and a DC voltage, from 100 to several thousand volts, is applied across the electrodes. Usually the solvent is an aqueous buffer of a pH that causes some components to be charged and thus attracted to one of the electrodes. If the sample components migrate to different electrodes or even to different degrees toward the same electrode, a separation is accomplished.

Dialysis

The two phases involved in dialysis are both liquids, but they are separated by a semipermeable membrane. As a result, the phases need not be immiscible; one is called the retentate and the other the diffusate. In operation the membrane controls the selectivity, not the liquid phases. The porosity of the membrane is so chosen that only some solutes can pass through it, and this selective permeation is the basis of separation.

Dialysis is usually slow and not too useful for routine analytical work. However, Craig (3) has described a thin-film dialysis column that is faster and somewhat analogous to LLC with a membrane (without a packed bed). This type of dialysis is described in Chapter 17.

Related to dialysis is the phenomenon called *osmosis*. The difference is that in osmosis the solvent (not the solute) passes through the membrane. Although this is an important process and one that probably occurs during dialysis, it is less useful in analytical separations.

CLASSIFICATION OF SEPARATION METHODS

There are many different ways of classifying separation methods. Each of them has a point of view that is worth examining as this helps to emphasize similarities and differences between methods. However, it must be remembered that many classification schemes are not exact and leave room for interpretation. A good summary of classification schemes was published by Elving in 1951 (4), but, as one would expect, chromatography was virtually excluded from consideration at that time. The classifications that follow include chromatography, but only the first one (Table 2) attempts to include all the major separation methods. The other classifications are less extensive.

Table 2.   Classification of Separation Methods by Type of Phase[a]

Initial Phase[b]	Second Phase		
	Gas (Vapor)	Liquid	Solid
Gas	Thermal diffusion	Gas–liquid chromatography	Gas–solid chromatography
Liquid	Distillation (Foam fractionation)	Liquid–liquid chromatography Liquid–liquid extraction Dialysis Ultrafiltration	Liquid–solid chromatography Precipitation Electrodeposition Crystallization Inclusion-compound formation (Zone electrophoresis) Ring oven
Solid	Sublimation	Zone refining Leaching	

[a] The classification of the methods enclosed in parentheses is less definitive than that of the others.
[b] Contains the sample or is the sample.

The Phases

As already indicated, most separation methods involve two phases. In Table 2, methods are classified according to the states of these two phases. The initial phase is taken as the one that contains the sample or is the sample. The methods that do not exactly fit in a given location in the table are enclosed in parentheses, and some others may be debatable. For example,

Table 3. Classification of Some Separation Methods by Type of Initial Phase

Sample Added to Initial Phase	Sample Is Initial Phase
Chromatography	Distillation
Liquid–liquid extraction	Crystallization
Zone electrophoresis	Zone refining
Differential dialysis	
Precipitation	

some may prefer to list distillation as a liquid–liquid method since the process starts with a liquid and ends with a liquid. However, the partitioning occurs between the starting liquid and the vapor formed, so it is classified as a liquid–gas (vapor) method.

The phase labeled "initial phase" in Table 2 either contains the sample or is the sample. These two alternatives give rise to the classification shown in Table 3. In those cases where the sample is the initial phase, fairly large quantities of sample are required for analysis.

Although the classification in Table 3 is a minor one, it does have an interesting resemblance to the next classification (Table 4), which is based on

Table 4. Classification of Some Separation Methods by Origin of Second Phase

Second Phase Added (Group I)	Second Phase Formed (Group II)
Chromatography	Distillation
Liquid–liquid extraction	Precipitation
Zone electrophoresis	Crystallization
Differential dialysis	Inclusion–compound formation
	Electrodeposition
	Zone refining
	Foam fractionation

the origin of the second phase. In group I, the second phase is *added*, whereas in group II the second phase is *formed*. The four methods in group I have several other features in common, which make this classification especially useful. The data obtained by these methods can be presented in graphical form as a series of peaks or bands like that shown in Figure 6 for a mixture of three solutes. The mechanisms by which the bands are broadened are treated in Chapter 8. The shapes of the bands approximate the normal Gaussian distribution, which will be discussed later in this chapter.

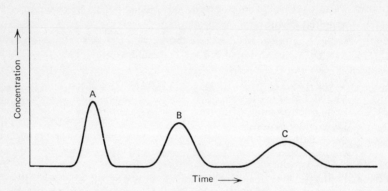

Figure 6.   Typical graph for a group I multistep separation.

Method of Contacting the Phases

There are two different schemes for classifying the method of contacting the phases. One is the distinction between continuous contact and stepwise or batch contact. Of all the techniques, extraction offers the most flexibility in this respect. Other techniques are usually limited to one method or the other; for example, chromatography by definition can be carried out only in a continuous manner.

The other classification is according to the direction of flow of the two phases relative to one another. Thus we have *co*current, *cross*current, and *counter*current. In a cocurrent process the initial phase (containing the sample) and a second phase are fed into a contacting unit so that they move through it in the same direction. As there are no common examples of this process in the analytical laboratory, it will not be discussed further.

In a crosscurrent process the phase containing the sample can be considered to be stationary. It is repeatedly contacted by successive portions of a second phase, which can be considered to "cross" it. Simple LLE with separatory funnels is a common example. In crosscurrent extraction a single raffinate phase is contacted with several volumes of extractant phase, which are then combined into a single extract. Other examples are simple (not fractional) distillation and repeated precipitation.

In a countercurrent process the two phases move in opposite directions relative to each other. Actually, in most cases one of the phases is stationary and the other one moves, so it is more properly called pseudo-countercurrent. This procedure differs from the crosscurrent process because successive portions of *both* phases are repeatedly added. Thus, in a stepwise process, each fraction of phase 1 is contacted by a fresh portion (no sample dissolved in it) of phase 2, and each fraction of phase 2 is contacted by a fresh portion of phase 1. Further discussion and a mathematical treatment of this process are given in Chapter 8. The common techniques carried out by the countercurrent method are LLE, chromatography, zone electrophoresis, countercurrent differential dialysis, and fractional distillation. It is obvious from this list that countercurrent processes are of primary interest in the analytical laboratory.

Type of Process

Separations can be classified according to the type of separation process: mechanical, physical, or chemical. Typical examples of each are shown in Table 5, although it must be recognized that these classifications are

Table 5. Classification of Separation Methods by Type of Process

Mechanical	Physical	Chemical
Sieving and exclusion (size):	Partition:	Changes of state:
Dialysis	Gas–liquid	Precipitation
Exclusion chromatography	chromatography	Electrodeposition
Inclusion–compound	Liquid–liquid	
formation	chromatography	
Filtration and	Gas–solid chromatography	
ultrafiltration	Liquid–liquid	
	chromatography	
	Liquid–liquid	Masking
	extraction	(pseudoseparation)
	Zone electrophoresis	
Centrifugation (density)	Foam fractionation	
		Ion exchange
	Changes of state:	
	Distillation	
	Sublimation	
	Crystallization	
	Zone refining	

approximate at best. Many techniques involve several of the three processes and therefore are not categorized as simply as the table would indicate. Most of our attention will be focused on the physical methods, which include chromatography; however, the traditional approach to quantitative analysis has emphasized the chemical methods, especially precipitation.

Included under chemical methods is masking, which is really not a separation method but rather a substitute for it. Since masking achieves the same goal as a separation, it is included in the table.

Driving Force

Some of the equations introduced earlier in this chapter apply to systems at equilibrium. The attainment of equilibrium can be considered to be a "driving force" for separations. The other type of driving force is kinetic, whereby the rate at which solutes travel determines the separation. Examples of these two types are given in Table 6. Note that chromatography is classed with the

Table 6.   Classification of Some Separation Methods by Type of Driving Force

Equilibrium	Kinetic
Chromatography	Dialysis
Distillation	Zone electrophoresis
Extraction	Diffusion, thermal
Zone refining	Centrifugation and ultracentrifugation
Sublimation	Ultrafiltration
Precipitation	

equilibrium methods, even though it was stated earlier that chromatographic systems are not at equilibrium. Nevertheless, the reason for chromatography's effectiveness is the tendency or driving force toward equilibrium. This classification is responsible in part for separate chapters on equilibrium (Chapter 3) and kinetics (Chapter 4).

Other Classifications

In the description of chromatography presented earlier the distinction was made between discontinuous (zonal or batch) and continuous sampling. This classification applies to other techniques as well. It is listed below along with several other minor classifications:

1. Sample introduction: zonal or continuous
2. Type of phase contact: bulk or interfacial
3. Nature of sample: aqueous or nonaqueous
4. Nature of sample: ionic or nonionic

These need no further elaboration but will be useful in the selection of a separation method (Chapter 18).

PRESENTATION OF DATA FOR GROUP I METHODS

When group I methods (see Table 4) are carried out in many steps or continuously, the resulting separation can be represented by graphs of concentration versus time (or some other related variable such as a number of steps, volume of one of the phases, etc.). Such a graph is shown in Figure 6 for a three-component mixture of A, B, and C. The straight line part of the graph between the three sample "peaks" is called the baseline. For a chromatographic process this graph is called a chromatogram.

Note that the peak for C is wider than the one for B, which is wider than the one for A. It will be shown later that this is a general phenomenon: that the longer time a solute spends in the system, the wider its peak becomes. The object of group I separation methods is to get the solutes separated by a distance greater than their peak widths in a minimum of time (see Chapter 9).

The peaks shown in Figure 6 follow a normal, or Gaussian, distribution. This is the ideal shape, but it is not always achieved in practice. Theoretically, the Gaussian shape is closely approached if the number of steps or stages is great enough. Unless otherwise noted, Gaussian peak shapes will be assumed in subsequent discussions.

The familiar equation for the normal distribution is

$$y = \frac{1}{\sigma\sqrt{2\pi}} \exp\left[-\frac{1}{2}\left(\frac{x - \bar{x}}{\sigma} \right)^2 \right] \qquad (27)$$

where y is the dependent variable, x is the independent variable, $\bar{x}$ is the average of a large number of x's, and σ is the standard deviation. In the analysis of peaks resulting from a separation it will be most useful to express the variable x in units of standard deviation. Hence, for our purposes, equation 27 can be rewritten as

$$y = \frac{1}{\sqrt{2\pi}} \exp\left(-\frac{\sigma^2}{2} \right) = 0.3989 \exp\left(-\frac{\sigma^2}{2} \right) \qquad (28)$$

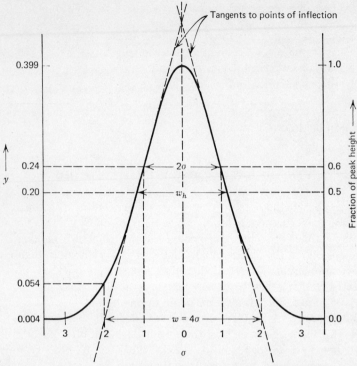

Figure 7. A normal distribution. The inflection point occurs at 0.607 of the peak height. The quantity w_h is the width at 0.500 of the peak height and corresponds to 2.354σ.

This equation is plotted in Figure 7. Also shown as broken lines are tangents to the points of inflection. Where they intersect the baseline, they cut off the distance w known as the peak width (at the base). It can be seen from the figure that w has a value of 4σ (±2σ). Consequently σ, the standard deviation, is also called the quarter peak width (at the base). Note also that the width at 60.7% of the peak height is 2σ (±1σ), and at 50% of the peak height it is 2.354σ. The latter is called the peak width at half height, w_h.

MEASURES OF SEPARATION EFFICIENCY

In concluding this chapter we need to consider the parameters that are used to evaluate or predict the efficiency of a separation. How can we tell if one separation is better than another? Some measures find use primarily in one technique and not in others, but a measure of separation efficiency that can be applied to a variety of techniques is obviously more valuable. As we consider the most common measures, this objective should be kept in mind. It should also be noted that expressions of the efficiency of a separation must refer to at least two solutes, whereas much of our discussion to this point has dealt with the behavior of only one solute.

Two types of expression can be distinguished. In one type the separation expression is concerned with only two solutes—either the two that are most difficult to separate or two that are designated as standards. This type will be discussed first, followed by the second type of expression, which specifies the efficiency of the system itself.

For Two Solutes

Separation Quotient.   The separation quotient is one of the most common expressions, but it is usually incorrectly called separation factor. In fact there is no common agreement on its name; some call it a separation ratio. We designate it by the symbol α and define it as

$$\alpha = \frac{(K_p)_B}{(K_p)_A} \tag{29}$$

It does not matter which solute, B or A, is in the numerator, but α is usually so defined that its numerical value is unity or greater. The larger the value of α, the easier it will be to achieve a good separation.

This definition is most easily applied to LLE. It says that for a good separation (large α) most of B will be left in the raffinate while most of A is extracted (refer to equation 6).

For other separation methods *equivalent* definitions are used. For example, in distillation, it is the ratio of vapor pressures p^0, taking A as the more volatile component:

$$\alpha = \frac{p_A^0}{p_B^0} \tag{30}$$

For chromatography it can also be expressed as the ratio of (adjusted) retention or elution volumes V_R', taking B as the component more strongly retained:*

$$\alpha = \frac{(V_R')_B}{(V_R')_A} \tag{31}$$

Actually α is less useful an expression than it may seem to be. The efficiency of a separation also depends on the actual values of the partition coefficients.

* The prime in the term V_R' has a specific and important meaning and designates an "adjusted" volume, but this will have to be deferred until Chapter 11.

For example, both of the following ratios calculate an α value of 4, but the first one represents a much better separation:

$$\alpha = \frac{2.0}{0.5} = 4.0 \tag{32a}$$

$$\alpha = \frac{20}{5} = 4 \tag{32b}$$

As a general rule, the separation is most efficient when the product of the two partition coefficients is unity (5). Consider, for example, an ideal extraction system in which one solute is extracted preferentially ($K_p < 1$) and the other solute remains largely in the raffinate ($K_p > 1$). This is an efficient system since the two solutes are in different solutions, and it is one in which the product of the K_p values will be close to unity. Obviously it is a better system than one in which both solutes are extracted or one in which both solutes remain in the raffinate. In both of these examples the solutes are present in the same solution, and the product of their K_p values would be far from unity.

For the examples in equations 32a and 32b,

$$(K_p)_A \times (K_p)_B = 2 \times 0.5 = 1.0 \tag{33a}$$
$$(K_p)_A \times (K_p)_B = 20 \times 5 = 100 \tag{33b}$$

In summary, the extractant in LLE should be so chosen that the partition coefficients of the solutes to be separated have a large quotient and a product near unity.

The relative volumes of the two phases can be varied to improve a separation (refer to equation 6). It has been shown that the best separations are obtained when the fraction of solute A in the raffinate equals the fraction of solute B in the extractant, or vice versa (6, 7). This condition prevails when

$$\beta = [(K_p)_A(K_p)_B]^{1/2} = \frac{V_{\text{extr}}}{V_{\text{raff}}} \tag{34}$$

Applying this rule to our previous examples, we find that for the example given in equation 32a, $\beta = 1^{1/2} = 1.0$, and for equation 32b, $\beta = 100^{1/2} = 10$. For the latter example this means that the volume of the extract phase should be 10 times as large as the volume of the raffinate phase for best results. This is not unexpected since both $(K_p)_A$ and $(K_p)_B$ are greater than unity, indicating that neither A nor B is easily extracted; that is, both A and B remain primarily in the raffinate, and a large volume of extractant is needed to get them out. Obviously, β can become too large or too small, so that the phase volumes are very dissimilar and difficult to work with.

In general we conclude that the separation quotient α has some value in expressing the ratio of basic parameters, but that by itself it is not an adequate measure of separation efficiency. Furthermore, there is some misconception and confusion over its name and use; it is not a separation *factor*.

Separation Factor. A true separation factor has been explained by Sandell (8) and given the symbol $S_{B/A}$. If A is the solute we want to isolate in as pure a form as possible, the appropriate separation factor is

$$S_{B/A} = \frac{R_B}{R_A} = \frac{\text{recovery factor for B}}{\text{recovery factor for A}} \tag{35}$$

Recovery factors represent the fraction of solute isolated:

$$R_A = \frac{W_A}{(W_A)_T} = \frac{\text{quantity of A isolated}}{\text{total quantity of A}} \tag{36}$$

In LLE, recovery factors are the same as Q, the fraction extracted. Hence for a one-step extraction equation 35 becomes

$$S_{B/A} = \frac{Q_B}{Q_A} \tag{37}$$

and if our earlier observations regarding the similarity between Q (in LLE) and R_R (in chromatography) are correct, the separation factor for chromatography is the ratio of R_R values.

As defined, $S_{B/A}$ is intended for use when A is nearly totally extracted, Q_A is nearly unity, and $S_{B/A}$ is approximately equal to the fraction of B extracted. The term $S_{B/A}$ is truly a factor, and if the original weight ratio of B/A is multiplied by $S_{B/A}$, the result is the weight ratio of B/A in the extract. Ideally, this should be a small number. Some people prefer to call $S_{B/A}$ an *enrichment factor* for A, but Sandell (8) argues that a better name is *depletion factor* for B.

For single-step LLE, the separation factor is

$$S_{B/A} = \frac{(K_p)_A + \beta}{(K_p)_B + \beta} \tag{38}$$

Thus we can see that $S_{B/A}$ is related to, but not exactly the same as, the reciprocal of the analogous separation quotient α.

Extent of Separation. The newest suggestion for a universal index of separation has been made by Rony (1), who calls it the extent of separation, ξ, which for a binary system is defined as

$$\xi = \text{abs det} \begin{vmatrix} (Y_A)_1 & (Y_A)_2 \\ (Y_B)_1 & (Y_B)_2 \end{vmatrix} \tag{39}$$

where Y is the segregation factor for solute i (A or B) in the region j (1 or 2)

$$(Y_i)_j = \frac{(N_i)_j}{(N_i)_T} \tag{40}$$

Here $(N_i)_j$ is the number of moles of i in region j, and $(N_i)_T$ is the total number of moles. Obviously Rony's segregation factor is the same as Sandell's recovery factor, which is the same as the fraction extracted Q in LLE:

$$(Y_A)_1 = Q_A; \qquad (Y_A)_2 = 1 - Q_A$$
$$(Y_B)_1 = Q_B; \qquad (Y_B)_2 = 1 - Q_B \tag{41}$$

In this case the determinant can be reduced to

$$\xi = \text{abs} \,|(Y_A)_1 - (Y_B)_1| = \text{abs} \,|Q_A - Q_B| \tag{42}$$

Thus for a single-step extraction, the extent of separation is the difference between the fractions extracted and the separation factor (equation 37) is the ratio of the fractions.

Rony's extent of separation has several advantages, including the fact that it is normalized; a good separation approaches the value 1.0, whereas a poor separation approaches zero. However, it has not yet gained wide acceptance.

Resolution. Resolution is a measure of separation efficiency that is also widely used in other contexts. Unlike the other parameters discussed, resolution is usually defined from a graph like that shown in Figure 8. (We have noted that this is the type of graph obtained for group I separations.) Resolution R_s is defined as

$$R_s = \frac{d}{(w_A + w_B)/2} = \frac{2d}{w_A + w_B} \tag{43}$$

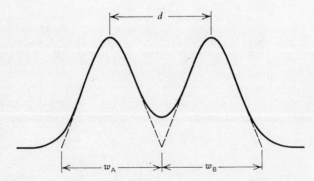

Figure 8. Two nearly resolved peaks illustrating the definition of resolution.

where d is the distance between the maxima of the peaks of the two solutes and w is the width of each peak at its base (peak width) as defined earlier.

In Figure 8 the tangents to the inflection points for solutes A and B are just touching; that is, w_A is touching w_B. Since the peaks are so close together, we can assume that their widths are the same. (Recall that peak widths increase as the time spent in the system increases.) Thus, if the width of each peak is 4σ, the sum of the two is 8σ. Similarly the distance of peak separation, d, can be stated in standard deviation units. Its value is 2σ from peak A plus 2σ from peak B, or 4σ. Hence the numerical value for R_s for two peaks whose tangents are just touching (98% resolved) is

$$R_s = \frac{2(4\sigma)}{8\sigma} = 1.0 \qquad (44)$$

A resolution of about 1.5 (equivalent to 6σ) is necessary for complete (baseline) separation, and values less than 1.0 represent poorer separations than that shown in Figure 8.

This definition is valid only when both peaks have the same height. A theoretical approach to the problem of unequal peaks in GC was published by Glueckauf (9) and is included in many GC textbooks (see, for example, reference 10), but it is not correct (11), and a more practical approach has been suggested recently by Snyder (12), whose paper includes large figures of various resolutions for a variety of peak heights, from 1/1 to 128/1. If direct measurements are preferred, the method of Carle (13) can be used.

Another modification of the basic equation has been suggested by Karger (14) for those situations when the adjacent peaks differ significantly in peak width. Finally, Bly (15) has suggested a modified equation for gel chromatography.

For the System

Number of Theoretical Plates and HETP.   The terms "theoretical plate" and HETP were originally used to express the efficiency of the liquid–vapor contacting process of a distillation column, but they have been carried over into chromatography. Precise definitions will be deferred until Chapters 7 and 11, but it can be noted that a good distillation column has a large number of theoretical plates, n. The acronym HETP stands for "height equivalent to a theoretical plate"; it is the height (or length) of the column that can be considered to contain one theoretical plate. Thus

$$H = \frac{L}{n} \qquad (45)$$

where L is the length of the distillation or chromatographic column. A good column will have a small HETP.

Number of Stages.   Some chromatographers do not like the concept of theoretical plates since the idea does not fit with the current theory of chromatography. Furthermore, it is not useful for comparing efficiencies in distillation and chromatography; equivalent separations require many more (at least 10 times as many) plates in chromatography than in distillation. This led Rony (*16*) to define a new term called *the number of stages*, which represents the number of stages in a countercurrent multistage column required to obtain a separation identical with that achieved in the chromatographic system.*
Using chromatographic symbols, Rony's number of stages can be calculated from equation 46, in which t_R and σ must be measured in the same units (usually distance on a chart):

$$\text{number of stages} = \frac{1}{\sqrt{2\pi}}\left(\frac{t_R}{\sigma}\right) \tag{46}$$

This equation can be compared with equation 5 in Chapter 11 for a comparison between the number of stages and the number of theoretical plates in chromatography. The number of stages has not received serious consideration by separation scientists and is not widely used.

Peak Capacity.   Another method of specifying the efficiency of a separation process, called the peak capacity (*17*), is based on the determination of the maximum number of solutes (peaks) that can be resolved (R_s of 1.0) by a given system in a given time (or total volume of mobile phase). Figure 9 shows a

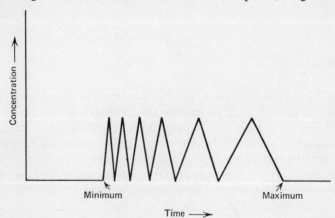

Figure 9.   A hypothetical separation illustrating the concept of peak capacity.

* Note error in equation 72 in Rony's paper (*16*).

series of solutes represented by triangles and comprising the maximum number that can be resolved in the total time indicated. Peak capacity is used in chromatography (*18*) and in electrophoresis and sedimentation (*19*).

A similar concept has been suggested by Kaiser (*20*), who calls his parameter a *separation number*. It is the number of solutes that can be resolved between two consecutive members of the paraffin homologous series. This and the other measures of separation efficiency used in chromatography have been fully compared by Ettre (*21*).

REFERENCES

1. P. R. Rony, *Separ. Sci.* **3**, 239 (1968).
2. J. C. Giddings, *Dynamics of Chromatography*, Part I, Dekker, New York 1965, p. 2.
3. L. C. Craig and H. Chen, *Anal. Chem.* **41**, 590 (1969).
4. P. J. Elving, *Anal. Chem.* **23**, 1202 (1951).
5. L. B. Rogers, *J. Chem. Educ.* **45**, 7 (1968).
6. M. T. Bush and P. M. Densen, *Anal. Chem.* **20**, 121 (1948).
7. E. Grushka, *Separ. Sci.* **7**, 293 (1972).
8. E. B. Sandell, *Anal. Chem.* **40**, 834 (1968).
9. E. Glueckauf, *Trans. Faraday Soc.* **51**, 34 (1955).
10. S. Dal Nogare and R. S. Juvet, *Gas–Liquid Chromatography*, Interscience, New York 1962, p. 83.
11. S. H. Tang and W. E. Harris, *Anal. Chem.* **45**, 1979 (1973).
12. L. R. Snyder, *J. Chromatogr. Sci.* **10**, 200 (1972).
13. G. C. Carle, *Anal. Chem.* **44**, 1905 (1972).
14. B. L. Karger, *J. Gas Chromatogr.* **5**, 161 (1967).
15. D. D. Bly, *J. Polymer Sci.*, Part C, **21**, 13 (1968).
16. P. R. Rony, *Separ. Sci.* **5**, 121 (1970).
17. J. C. Giddings, *Anal. Chem.* **39**, 1027 (1967).
18. E. Grushka, *Anal. Chem.* **42**, 1142 (1970).
19. J. C. Giddings, *Separ. Sci.* **4**, 181 (1969).
20. R. Kaiser, *Chromatographie in der Gasphase*, Vol. II, 2nd ed., Bibliographisches Institut, Mannheim, Germany, 1966, pp. 47–48.
21. L. S. Ettre, *J. Gas Chromatogr.* **1**[2], 37 (1963).

SELECTED BIBLIOGRAPHY

General

Anal. Chim. Acta **38**, 1–315 (1967).
Berg, E. W., *Physical and Chemical Methods of Separation*, McGraw-Hill, New York, 1963.
Danatos, S., *Chem. Eng.* **71**(25), 155 (1964).

Dean, J. A., *Chemical Separation Methods*, Van Nostrand–Reinhold, New York, 1969.

Karger, B. L., L. R. Snyder, and C. Horvath, *Introduction to Separation Science*, Wiley-Interscience, New York, 1973.

Kolthoff, I. M., and P. J. Elving (eds.), *Treatise on Analytical Chemistry*, Vols. 2 and 3, Part I, Wiley-Interscience, New York, 1961.

Li, N. N. (ed.), *Recent Developments in Separation Science*, 2 vols., CRC Press, Cleveland Ohio, 1972.

Perry, E. S., and C. J. van Oss (eds.), *Progress in Separation and Purification*, 4 vols., Wiley-Interscience, New York, 1968.

Perry, E. S., and C. J. van Oss (eds.), *Separation and Purification Methods*, Vol. 1, Dekker, New York, 1973.

Schoen, H. M., and J. J. McKetta, Jr., *New Chemical Engineering Separation Techniques* Wiley-Interscience, New York, 1962.

Weissberger, A. (ed.), *Technique of Organic Chemistry*, Vol. III, Part 1, Interscience, New York, 1956.

Welcher, F. (ed.), *Standard Methods of Chemical Analysis*, 6th ed., Vol. 2, Part A. D. Van Nostrand, Princeton, N.J., 1963.

Wilson, C. L., and D. W. Wilson (eds.), *Comprehensive Analytical Chemistry*, American Elsevier, New York, 1959.

Zone Refining

Pfann, W. G., *Zone Melting*, 2nd ed., Wiley-Interscience, New York, 1966.

Bollen, N. J. G., M. J. Van Essen, and W. M. Smit, *Anal. Chim. Acta* **38**, 279 (1967).

Schildknecht, H., *Anal. Chim. Acta* **38**, 261 (1967).

Ring Oven Technique

Weisz, H., *Microanalysis by the Ring Oven Technique*, Pergamon Press, London, 1960.

Berg, E. W., *Physical and Chemical Methods of Separation*, McGraw-Hill, New York, 1963, Chapter 8.

Bubble Separation Methods (Including Foam Fractionation)

Cassidy, H. G., *Fundamentals of Chromatography*, Vol. X in the series, *Technique of Organic Chemistry*, Weissberger, ed., Interscience, New York, 1957, p. 327.

Garmendia, A. A., D. L. Pérez, and M. Katz, *J. Chem. Educ.* **50**, 864 (1973).

Karger, B. L., and P. G. DeVivo, *Separ. Sci.* **3**, 393 (1968).

Lemlich, R., *Adsorptive Bubble Separation Techniques*, Academic Press, New York, 1972.

HETEROGENEOUS EQUILIBRIUM AND THERMODYNAMICS

3

Pfann (*1*) has written an interesting keynote article for the journal *Separation Science*. In his survey of this new field he notes that "separation is against Nature" and "thermodynamics seems to be against purification of materials." He is, of course, referring to the second law of thermodynamics, on the basis of which one must conclude that a separation is work.

The purpose of this chapter is to examine the relationships between classical thermodynamics and those separation methods that were classified as equilibrium methods in Chapter 2 (Table 6). Thermodynamics is the most common way of describing equilibrium systems, and the equations and figures presented in this chapter provide a foundation for later discussions. However, this brief summary should be supplemented with additional study in thermodynamics; some suggested readings are listed at the end of the chapter.

BASIC DEFINITIONS AND EQUATIONS

Equilibrium can be defined as the condition that exists when the macroscopic properties of a system do not change with time. For example, in a liquid–liquid extraction, equilibrium has been reached when further shaking of the separatory funnel does not result in any additional extraction. This is not to deny that the equilibrium is dynamic and that solute is continually passing from one phase to the other. Rather, the rate of change is the same in both directions.

Thermodynamically, equilibrium exists when there is no further change in the free energy $\mathscr{G}$ of a system at constant temperature T and pressure P:

$$(\Delta \mathscr{G})_{T,P} = 0 \tag{1}$$

Since $\mathscr{G} = \mathscr{H} - T\mathscr{S}$, the free-energy change for an isothermal process is

$$\Delta \mathscr{G} = \Delta \mathscr{H} - T\Delta \mathscr{S} \tag{2}$$

where $\mathscr{H}$ is enthalpy and $\mathscr{S}$ is entropy. Although we will make little direct use of equation 2, it is useful to remember that the free-energy change of a process depends on both enthalpy and entropy changes.

In addition to the Gibbs free energy $\mathscr{G}$, equilibrium can be described by the chemical potential χ. It is one of the so-called partial molal quantities (2), which can be defined as the rate of change in the content of a particular extensive quantity of a system while a particular component is added.* Specifically, the chemical potential of substance i, χ_i, is the rate of change in *free energy* of the system with the number of moles of i, the temperature, pressure, and number of moles of other components being kept constant:

$$\chi_i = \left(\frac{\partial \mathscr{G}}{\partial N_i}\right)_{T,P,N_j} \tag{3}$$

Thus the chemical potential is just the partial molal free energy $\overline{\mathscr{G}}$. The sum of the partial molal free energies of a system equals the free-energy change of the system:

$$\Delta \mathscr{G} = \sum_i \overline{\mathscr{G}}_i \Delta N_i \tag{4}$$

In these terms equilibrium is defined as the condition in which the chemical potential of a component is equal in each of two phases. For example, if A is being partitioned between phases 1 and 2, equilibrium can be expressed as

$$(\chi_A)_1 = (\chi_A)_2 \tag{5}$$

Since absolute values of χ or $\mathscr{G}$ cannot be measured, standard states are arbitrarily defined (indicated by superscript zero), and changes in these properties are defined according to equation 6:

$$\chi = \chi^0 + \mathscr{R}T \ln \frac{p}{p^0} \tag{6}$$

where $\mathscr{R}$ is the gas constant and p and p^0 are vapor pressures.† The standard-state vapor pressure p^0 is defined as 1 atm, so equation 6 reduces to

$$\chi = \chi^0 + \mathscr{R}T \ln p \tag{7}$$

The choice and specification of standard states must be done with care to avoid pitfalls such as those described for GC systems (3).

A relation similar to equation 7 can be stated in terms of the standard free-energy change at constant pressure, $\Delta \mathscr{G}°$; for systems at equilibrium:

$$\Delta \mathscr{G}° = -\mathscr{R}T \ln K_p \tag{8}$$

---

* It is important to distinguish between the extensive quantities of a system (such as $\mathscr{G}$, $\mathscr{H}$, and $\mathscr{S}$) and the intensive quantities (such as P, T, V, and χ). The latter are uniform throughout a system, whereas the former vary with the size of the system. Thus each component of a solution makes a contribution to the total of each extensive property of the solution.

† The expression "ln" is used to denote the logarithm to base e, and "log" is used to denote the logarithm to base 10.

This equation relates the fundamental thermodynamic property free energy to the one used to describe equilibrium systems, the partition coefficient K_p.

One final relationship is commonly known as the Clapeyron equation:

$$\frac{dp}{dT} = \frac{\Delta \mathscr{H}}{T \Delta V} \tag{9}$$

It describes the change in pressure p that occurs with a change in temperature T for any change of state (e.g., melting, evaporating). The quantity ΔV is the difference in molar volume between the two states, and $\Delta \mathscr{H}$ is the enthalpy change associated with the change in state. A plot of equation 9 for a given substance is called a phase diagram, and this will be discussed later.

For evaporation (liquid-to-vapor change in state), ΔV can be replaced by the partial molal volume of the vapor since the volume of the liquid will be small by comparison. According to the ideal gas law the volume can be replaced by $\mathscr{R}T/P$. When these substitutions are made, equation 9 becomes

$$\frac{d(\ln P)}{dT} = \frac{\Delta \mathscr{H}}{\mathscr{R}T^2} \tag{10}$$

which is known as the Clausius–Clapeyron equation and is applicable to liquid–vapor systems. If $\Delta \mathscr{H}$ is assumed to be constant, this equation can be integrated to give the following two equations, depending on whether or not the integration is between limits:

$$\log p^0 = -\frac{\Delta \mathscr{H}}{2.3\mathscr{R}T} + \text{constant} \tag{11a}$$

$$\log \frac{(p^0)_a}{(p^0)_b} = -\frac{\Delta \mathscr{H}}{2.3\mathscr{R}}\left(\frac{1}{T_a} - \frac{1}{T_b}\right) = -\frac{\Delta \mathscr{H}}{2.3\mathscr{R}}\left(\frac{T_b - T_a}{T_a T_b}\right) \tag{11b}$$

In equation 11b, p_a^0 and p_b^0 represent the vapor pressures of a given compound at two temperatures, T_a and T_b.

These equations are important in distillation and gas–liquid chromatography. For example, a plot of equation 11a for pentane is shown in Figure 1. Since the slope of this straight line is $-(\Delta \mathscr{H}/2.3\mathscr{R})$, this plot can be used to determine the $\Delta \mathscr{H}$ value. Similar plots for GC can be related to retention parameters.

PARTITION EQUILIBRIA

In the techniques of extraction and chromatography the equilibrium condition will usually be expressed in terms of a partition coefficient, K_p. In that

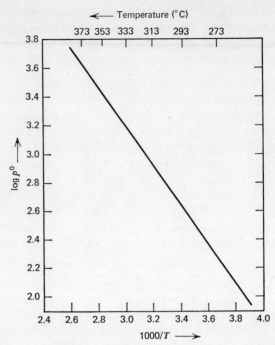

Figure 1. Variation of vapor pressure of pentane with temperature. A plot of the Clausius–Clapeyron equation.

case equation 8 can be used to state the relationship to the fundamental thermodynamic property free energy.

Equations 8 and 11 will be used in Chapter 12 to describe the temperature dependence of GC retention volumes.

CHANGE OF STATE

Distillation

Some of the discussions about distillation will be based on the Clausius–Clapeyron equation (equations 11a and 11b). It has already been noted that it can be used to determine values of $\Delta\mathscr{H}$ for changes of state, including evaporation in the distillation process.

Empirically it has been found that the value for $\Delta\mathscr{H}_{vap}$, the heat of vaporization of a liquid, can be estimated by Trouton's rule:

$$\Delta\mathscr{H}_{vap} = 21 T_{boil} \tag{12}$$

where T_{boil} is the boiling point of the liquid in degrees Kelvin. So if the boiling point of a liquid is known, its $\Delta\mathscr{H}_{vap}$ can be estimated. Then, using equation 11 as well as the fact that $(p^0)_b$ is 760 torr when T_b is the normal boiling point, the boiling temperature at some other pressure $(p^0)_a$ can be found.

Example: The boiling point of *n*-decane is 174°C. What is its boiling point in a vacuum distillation at 10-torr pressure?

$$\Delta\mathscr{H}_{vap} = 21 \times 447°K = 9.4 \times 10^3 \text{ cal}$$

$$\log p^0 - \log 760 = -\frac{9.4 \times 10^3}{2.3 \times 2}\left(\frac{1}{T_a} - \frac{1}{447}\right)$$

$$T_a = 318°K = 45°C$$

The experimental value is 58°C. For convenience in making this type of calculation, nomographs have been prepared (*4*).

Furthermore, if it is assumed that two substances have the same $\Delta\mathscr{H}$, equation 11 can be rewritten as a comparison of *two* substances whose vapor pressures are p_A^0 and p_B^0, and whose boiling points are T_A and T_B:

$$\log\frac{p_A^0}{p_B^0} = -\frac{\Delta\mathscr{H}}{2.3\mathscr{R}}\left(\frac{1}{T_A} - \frac{1}{T_B}\right) \tag{13}$$

Rose (*5*) has derived another equation useful in distillation by combining Trouton's rule and the Clausius–Clapeyron equation:

$$\log\frac{p_A^0}{p_B^0} = 8.9\frac{T_B - T_A}{T_A + T_B} \tag{14}$$

where $T_B > T_A$, which is to say component A is more volatile than B. Since $p_A^0/p_B^0 = \alpha$, as defined in Chapter 2,

$$\log\alpha = 8.9\frac{T_B - T_A}{T_A + T_B} \tag{15}$$

Precipitation

Precipitation is the other important separation technique that involves a change in state. Unlike all of the other methods mentioned, it involves a chemical reaction. The equilibrium condition is described by the solubility product (or ion product) K_{sp}, which is another equilibrium constant. For the hypothetical precipitation reaction

$$a\text{M} + b\text{X} \rightleftharpoons \underline{\text{M}_a\text{X}_b} \tag{16}$$

the K_{sp} expression is

$$K_{sp} = [M]^a[X]^b \tag{17}$$

again assuming ideal behavior. It is assumed that calculations of solubility have already been covered adequately.

Phase Rule

The phase rule derived by Gibbs in 1876 should also be used in discussing changes in state (6). It is

$$g = c - b + 2 \tag{18}$$

where b is the number of phases, c is the number of components, and g is the number of "degrees of freedom." The meaning of "number of phases" should be clear, but the other two terms need further definition. The number of components is the minimum number required to express the composition of the system. This means that any relationships between components (such as chemical reactions) will make c equal to the total number of substances *minus* the number of relationships between them. Ordinarily we will not be considering any chemical reactions, and the number of components should be obvious. For most of our examples, $c = 2$, as, for example, in a distillation of a binary mixture.

The number of degrees of freedom, g, is the number of independent, intensive variables needed to describe the system completely. The three

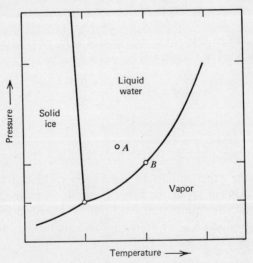

Figure 2. Phase diagram for water (not to scale).

intensive variables of interest in separations are temperature, pressure, and composition (concentration). By specifying two of the three, the third is automatically determined; this is the reason for the constant 2 in equation 18.

The phase rule, in effect, determines the nature of so-called phase diagrams, which are useful in understanding separations. A few examples will be considered, the simplest of which is a one-component diagram like Figure 2, the phase diagram for water. Since the number of components is one, $g = 3 - b$. For point A in the liquid area of the diagram ($b = 1$), $g = 2$. Thus both temperature and pressure must be specified to locate A. To locate point B, which is on the vapor–liquid equilibrium line ($b = 2$), only one variable must be specified, either temperature or pressure.

Next consider a system containing two components ($c = 2$), like benzene and toluene in Figure 3. In plotting the phase diagram, one of the three intensive variables must be kept constant and the other two plotted if we are to have a two-dimensional diagram. In Figure 3a, P is constant and T is plotted versus composition (mole fraction). As in Figure 2, point A is in the liquid region of the diagram ($b = 1$) and $g = 4 - 1 = 3$. All three variables must be known to locate A.

Point B is on a vapor–liquid equilibrium line ($b = 2$) and so $g = 2$. Pressure and either temperature or composition must be specified to locate B. Table 1 lists the possible values for g and b when $c = 2$.

Figure 3a is one type of phase diagram useful in describing distillation. From it one can see that the boiling point of a 50:50 mixture (mole basis) of benzene and toluene (at 1-atm pressure) is 92°C. Furthermore the vapor in equilibrium with this liquid mixture is composed of 78 mole % benzene. As expected, the vapor is richer in the more volatile substance.

In Figure 3b the pressure is plotted versus composition the temperature being held constant. The broken lines show the partial vapor pressure of each of the two components. The equation for these lines is known as Raoult's law:

$$p_A = X_A p_A^0 \tag{19}$$

where p_A is the partial vapor pressure of substance A, p_A^0 is the vapor pressure of pure A, and X_A is the mole fraction of A. As a solution of component A is diluted with component B and the mole fraction of A is decreased, the partial pressure of A is decreased since the proportion of molecules of A at the surface is decreased. No interactions are present between A and B, so solutions that follow Raoult's law are called ideal solutions.

The top (solid) line representing the boundary between liquid and liquid mixed with vapor is clearly the sum of the two broken lines. This is an expression of Dalton's law: the total pressure is equal to the sum of the individual partial pressures.

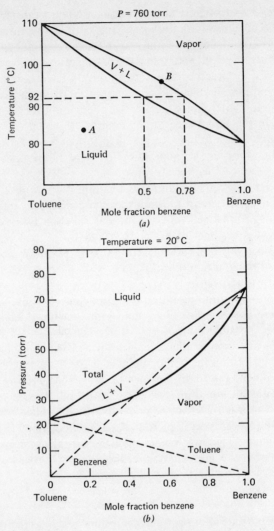

Figure 3. Phase diagram for the two-component mixture of benzene and toluene: (a) isobaric; (b) isothermal.

Table 1. Phase Rule
for $c = 2$

b	g
1	3
2	2
3	1
4	0

This discussion has centered on vapor–liquid equilibria, which are relevant to distillation, but diagrams similar to Figure 3a can also be drawn for liquid–solid equilibria. They are useful in explaining zone refining.

DEVIATIONS FROM IDEALITY

The examples used so far in this chapter were chosen because they were ideal systems and followed the laws under discussion. Actually, few mixtures are ideal and most deviate from Raoult's law. Several of these are shown in Figures 4, 5, and 6. The broken lines represent Raoult's law, which these substances do *not* follow. The solid lines represent the actual data. The deviations are classed as either positive (Fig. 4) or negative (Fig. 5) deviations from Raoult's law. These deviations result from intermolecular attractions and repulsions that cannot be described or predicted quantitatively at present. Positive deviations result when the attraction between A and B (acetone and carbon disulfide) is weaker than that for A–A and B–B. The opposite is true for negative deviations. The composition–temperature phase diagrams for nonideal solutions exhibit maxima or minima (Figs. 4 and 5), corresponding to compositions called azeotropes or azeotropic mixtures. They have serious consequences in distillation as we shall see later. When azeotropic solutions are boiled the vapor has the same concentration as the liquid and no separation occurs. As already noted, similar diagrams exist for liquid–solid equilibria.

Figure 6 shows a slightly different situation for a solid–liquid phase diagram. In this case naphthalene and camphor do not form a single solid solution. A mixture called an eutectic mixture, forms preferentially. Three phases are in equilibrium at that point: solid camphor solid naphthalene, and a liquid solution composed of 42 mole % naphthalene. Thus $g = 4 - 3 = 1$. The variable that must be specified is the pressure which is constant in this example.

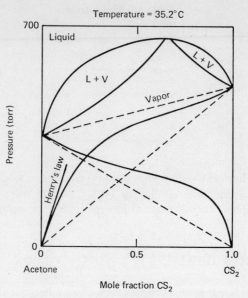

Figure 4. Phase diagram for the two-component system of acetone and carbon disulfide

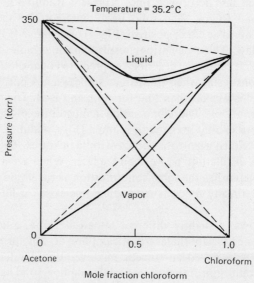

Figure 5. Phase diagram for the two-component system of acetone and chloroform.

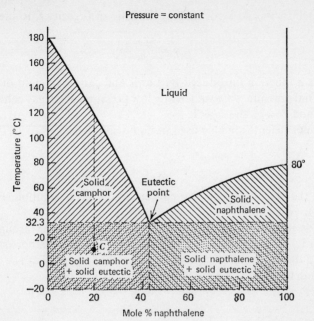

Pressure = constant

Figure 6. Phase diagram for the two-component system of naphthalene and camphor.

If a mixture of solids of composition C is heated it will begin to melt at 32.3°C to give a liquid with the composition of the eutectic. The temperature will remain constant during further heating until all of the naphthalene is melted and the system contains solid camphor and liquid eutectic. The camphor will melt on further heating. This type of system is especially amenable to purification (separation) by zone refining.

One final relationship between liquids and their vapors should be mentioned. If a nonideal solution is dilute enough a straight-line relationship exists over a short range of composition. This is shown in Figure 4. The equation for this line is known as Henry's law:

$$p_A = X_A k \qquad (20)$$

where k is a proportionality constant (rather than p_A^0 as in Raoult's law). Henry's law is stated for the solute (carbon disulfide in Figure 4 since its mole fraction is small) rather than the solvent. Most dilute solutions obey Henry's law over a limited range. Extensive use is made of this equation in describing dilute solutions of gas in a liquid.

Correction factors have been proposed for use with nonideal systems. For example, activity coefficients γ are used to correct Raoult's law:

$$p_A = \gamma_A X_A p_A^0 \qquad (21)$$

Activity coefficients can be defined as the ratio of fugacity f, to pressure:

$$\gamma = \frac{f}{p^0} \tag{22}$$

Fugacity is a function introduced by Lewis and can be thought of as an effective partial pressure. At a sufficiently low pressure, every gas behaves as an ideal gas, and $f = p^0$ and $\gamma = 1$.

Activity coefficients are also used in liquid solutions:

$$\gamma = \frac{a}{C} \tag{23}$$

where a is the activity or effective concentration of the solution. For dilute solutions, $a = C$ and $\gamma = 1$.

As already stated, ideality will be assumed for most systems throughout this book in order to keep the discussion simple. Standard textbooks on equilibrium and thermodynamics (see the Selected Bibliography) can be consulted for further detail.

REFERENCES

1. W. G. Pfann, *Separ. Sci.* **1**, 1 (1966).
2. See, for example, F. Daniels and R. A. Alberty, *Physical Chemistry*, 3rd ed., Wiley, New York, 1966, pp. 101 and 104.
3. M. R. James, J. C. Giddings, and R. A. Keller, *J. Gas Chromatogr.* **3**, 57 (1965).
4. S. B. Lippincott and M. M. Lyman, *Ind. Eng. Chem.* **38**, 320 (1946).
5. A. Rose, *Ind. Eng. Chem.* **33**, 594 (1941).
6. L. O. Case, in *Treatise on Analytical Chemistry*, Part I, Vol. 2, I. M. Kolthoff and P. J. Elving (eds.), Wiley-Interscience, New York, 1961, Chapter 23.

SELECTED BIBLIOGRAPHY

Andrews, F. C., *Thermodynamics: Principles and Applications*, Wiley-Interscience, New York, 1971.

Barrow, G., *Physical Chemistry*, 3rd ed., McGraw-Hill, New York, 1973.

Castellan, G. W., *Physical Chemistry*, 2nd ed., Addison-Wesley, Reading, Mass., 1971.

Daniels, F., and R. A. Alberty, *Physical Chemistry*, 3rd ed., Wiley, New York, 1966.

Klotz, I. N., and R. M. Rosenberg, *Chemical Thermodynamics*, 3rd ed., Benjamin, New York, 1972.

Moore, W. J., *Physical Chemistry*, 4th ed., Prentice-Hall, Englewood Cliffs, N.J., 1973.

Lewis, G. N., and M. Randall, *Thermodynamics*, 2nd ed., McGraw-Hill, New York, 1961.

KINETICS

4

Dialysis and zone electrophoresis are two of the techniques that were classified in Chapter 2 as being kinetically controlled. In dialysis the driving force is a concentration gradient and the discriminating element is a selective membrane. In electrophoresis the driving force is a voltage gradient and selectivity depends on the polarity of the ions and their mobilities.

Chromatography was discussed in the preceding chapter as an equilibrium technique since its retention parameters can be related to the thermodynamic partition coefficient. However, chromatography is dynamic, and concentration gradients exist in a solute zone as it passes through the chromatographic bed. Those gradients result in diffusion, a kinetic process treated in this chapter. Therefore a significant part of this chapter deals with the nature of a chromatographic bed and diffusion and mass transfer in it.

So far two types of gradient have been mentioned in connection with the kinetic separation methods: concentration and voltage. The nonequilibrium condition that gives rise to a kinetic process is a *gradient*. The common types of gradient are listed in Table 1 along with the separation processes controlled by them. We are concerned with concentration and voltage gradients only, and begin with a general discussion of concentration gradients and diffusion.

LAWS OF DIFFUSION

Diffusion can be defined as the process of migration of a species in a gradient. For the present we restrict our consideration to a concentration gradient, and later we shall consider a voltage gradient.

Consider two adjacent zones that have different concentrations of the same solute. Diffusion is the spontaneous process that results in the transfer of solute from the higher concentration to the lower one. This dilution and mixing are the result of the second law of thermodynamics, as was noted in Chapter 3. The individual molecules are moving as a consequence of their thermal agitation, and not in an individual attempt to move to a less populated region (1).

Table 1. Classification of Some Kinetic Separation Methods by Type of Gradient

Gradient	Process in Which It Operates
Concentration:	Diffusion:
No membrane	Chromatography
Membrane	Dialysis
Voltage	Electrophoresis
Gravity	Sedimentation and centrifugation
Temperature	Thermal diffusion

The main rules used to describe diffusion are Fick's two laws (*1, 2*). If $\partial C/\partial z$ is taken as the concentration gradient in the z-direction, Fick's first law can be expressed as

$$J = -D\left(\frac{\partial C}{\partial z}\right) \tag{1}$$

where J is the flux of mass transport through a unit area (perpendicular to z) per unit time and D is the diffusion coefficient. The units for D are square centimeter per second, and some typical values are given in Table 2. The minus sign in equation 1 indicates that the diffusion occurs from a higher to a lower concentration.

Fick's second law describes a change in concentration with time, $\partial C/\partial t$;

$$\frac{\partial C}{\partial t} = \frac{\partial}{\partial z}\, D\, \frac{\partial C}{\partial z} \tag{2}$$

which, if D is a constant, becomes

$$\frac{\partial C}{\partial t} = D\, \frac{\partial^2 C}{\partial z^2} \tag{3}$$

Table 2. Typical Diffusion Coefficients

System	D (cm²/sec)	Temperature (°C)
Liquids:		
Glucose–water	0.52×10^{-5}	15
CCl$_4$–methanol	1.7×10^{-5}	15
Acetic acid–benzene	1.92×10^{-5}	14
Gases:		
Argon–n-octane	0.059	30
Nitrogen–n-octane	0.073	30
Helium–n-hexane	0.574	144
Helium–methanol	1.032	150

Fick's first law will be used to describe diffusion in dialysis. His second law can be solved to yield the shape of a chromatographic zone. Assuming that the initial zone (sample) in chromatography is a sharp spike, integration of equation 3 yields an equation for a Gaussian distribution (1):

$$C = \frac{Z}{t^{1/2}} \exp\left(-\frac{z^2}{4Dt}\right) \tag{4}$$

where Z is a constant proportional to the total mass of diffusing sample. In effect, this equation says that a sample that starts out on a sharp spike

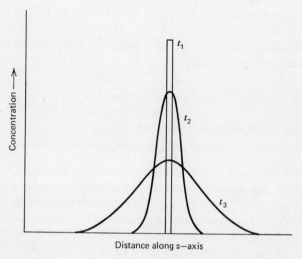

Figure 1. Zone widening due to diffusion. Three times are shown with $t_3 > t_2 > t_1$.

(narrow zone) will spread out in proportion to the square root of the time spent in the system—that is, in proportion to the square root of the distance migrated in the bed (at a constant flow rate). As it widens due to diffusion, the zone assumes a Gaussian shape, as shown in Figure 1. Further explanations follow in this chapter and in Chapter 8.

It is more convenient to write relations like equation 4 in terms of the standard deviation σ or quarter-band width of the Gaussian distribution. Since

$$C \propto \exp\left(-\frac{z^2}{2\sigma^2}\right) \tag{5}$$

we find that

$$\sigma^2 = 2Dt \tag{6}$$

which is called Einstein's relationship. The term D is the diffusion coefficient, as in earlier equations. We shall return to equation 6 later in describing chromatographic diffusion.

DIFFUSION THROUGH A MEMBRANE

In dialysis a concentration gradient is maintained between two solutions by separating them by a semipermeable membrane, and Fick's first law is usually rewritten with the following substitutions: the diffusion coefficient D is replaced by the product of k, a permeability constant, and A, the area of the membrane:

$$D = kA \tag{7}$$

The concentration gradient can be replaced by the difference between the two specific concentrations, C_0 and C_i:

$$\frac{\partial C}{\partial z} = C_0 - C_i \tag{8}$$

On integration, Fick's first law becomes

$$N = kA(C_0 - C_i)t \tag{9}$$

This equation shows that the amount of material diffused, N, is directly proportional to the area A, the concentration difference $(C_0 - C_i)$, and the time t. The permeability coefficient k is a characteristic of the membrane. Equation 9 is valid for a constant temperature. Furthermore, it assumes that the substance in question is small enough to pass through the pores of the membrane and that it is nonionic.

Donnan Equilibrium Theory

With regard to the previous discussion, it is interesting to observe the consequences of diffusion of ionic substances when one of them is too large to pass through the membrane. Consider two solutions, Na^+Cl^- and Na^+R^-, separated by a membrane whose pores are of such a size as to pass Na^+ and Cl^- ions, but not the large (hypothetical) R^- ions. In time, the Na^+ and Cl^- ions will distribute themselves equally on both sides of the membrane:

$$[Na^+]_1[Cl^-]_1 = [Na^+]_2[Cl^-]_2 \tag{10}$$

Furthermore, according to the electroneutrality principle,

$$[Na^+]_1 = [Cl^-]_1 \tag{11}$$

but

$$[Na^+]_2 = [Cl^-]_2 + [R^-]_2 \tag{12}$$

Thus

$$[Cl^-]_1^2 = [Cl^-]_2([Cl^-]_2 + [R^-]_2) \tag{13}$$

$$= [Cl^-]_2^2 + [Cl^-]_2[R^-]_2 \tag{14}$$

So

$$[Cl^-]_1 > [Cl^-]_2 \tag{15}$$

This is called the Donnan equilibrium theory, which is helpful in studying those cases where there is a nondiffusible electrolyte (like R^-). Thus at equilibrium the concentration of the Cl^- ion will differ on the two sides of the membrane. At first glance this would seem to contradict the concept of equilibrium, but of course it does not, as just shown.

Table 3. Calculations of the Fraction of Chloride Ions Transported Across a Membrane According to the Donnan Theory

$[Cl^-]_1/[R^-]_2$	$x/[Cl^-]_1$
$\dfrac{1}{100}$	0.01
$\dfrac{1}{100}$	0.33
1	$\sim$0.50

Another consequence of this phenomenon can be seen by deriving another equation, letting x equal the net amount of Cl^- ions diffused from side 1 to side 2:

$$\frac{x}{[Cl^-]_1} = \frac{[Cl^-]_1}{2[Cl^-]_1 + [R^-]_2} \tag{16}$$

Thus $x/[Cl^-]_1$ is the fraction of the diffusible Cl^- ion that has gone from side 1 to side 2 at equilibrium. Table 3 indicates that this fraction varies with the original ratio of $[Cl^-]_1/[R^-]_2$. It shows that by keeping the concentration of the nondiffusible ion high in phase 2 we can prevent the diffusion of the diffusible ion into that phase.

Donnan Equilibrium and Ion-Exchange Resins

The Donnan theory has been applied to ion exchange as well as to dialysis. An ion-exchange resin does not have any membrane associated with it, but since the resin can be considered to be a nondiffusible ion, R^- (e.g., for an anion resin), the Donnan equilibrium theory is applicable. In effect, the resin surface acts as a membrane separating the bulk solution from the solution in the pores of the resin.

It is this concept that helps us to explain why ions from a dilute solution are excluded from the interior of a high-capacity resin where the concentration of ionic sites is much higher. Further discussion of ion-exchange phenomena can be found in Chapter 6.

DIFFUSION AND MASS TRANSFER IN CHROMATOGRAPHY

Before discussing diffusion in chromatography, we need to have a better concept of the nature of a packed chromatographic bed. In the ideal case small solid particles of uniform diameter are tightly packed together. This solid can be the stationary phase itself (GSC and LSC) or it can be the solid support for a liquid stationary phase. If the solid particles are porous, which is often the case, then the solid contains interior pores that are accessible to the mobile phase. Some mobile phase can become trapped in these pores and be held immobile. In that case the stagnant mobile phase should be considered to be part of the stationary phase, which means that the nature and composition of the stationary phase can be very complex.

The total volume of a packed bed, V_T, can be divided into several parts:

V_{ss} = volume occupied by the solid support
V_S = volume occupied by the stationary phase
V_M = volume occupied by the mobile phase

For nonporous particles, V_M is the space between the particles, but for a porous particle (spongelike), V_M includs both the interparticle volume and the internal volume of the particles.

$$V_T = V_{ss} + V_S + V_M \tag{17}$$

The definition of V_M indicates that it is possible to define two porosities. The total porosity is the fraction of "dead space" in the bed

$$\text{total porosity} = \frac{V_M}{V_T} = \epsilon_T \tag{18}$$

It can be as high as 0.80 to 0.84 for a porous solid, indicating that even a tightly packed bed of porous material is largely dead space (3).

The other porosity term, ϵ_I, is only that fraction of volume between the particles:

$$\epsilon_I = \frac{\text{interparticle volume}}{V_T} \tag{19}$$

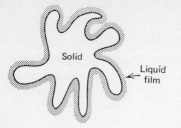
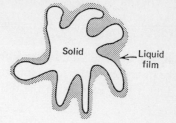

Figure 2. Uniform distribution of liquid phase on solid support.

Figure 3. Nonuniform distribution of liquid phase on solid support.

It varies with the tightness of the packing, with a typical value of 0.40 to 0.45 (*4, 5*). Nonporous, solid-core solids have only one porosity, ϵ, of course.

If the stationary phase is not a solid but a liquid, it must be held immobile by an inert solid support that is coated with a thin film of the liquid. Ideally this should be a uniformly thin coat, as shown in Figure 2, but due to capillary action, small pools of liquid occur (Fig. 3). Such pools can also form between the particles, as shown in Figure 4. Both types of pools are undesirable and are more likely to occur when large amounts of stationary liquid are used. Unfortunately the nature of the liquid phase is quite difficult to describe accurately.

Now that the nature of the chromatographic bed has been described, let us consider what happens to a single sample component when it is carried into the bed by the mobile phase. First consider a component that is not retarded at all by the stationary phase (i.e., it does not sorb on the stationary phase, or $K_p = 0$). Ideally it should enter the bed as a sharp spike, and hence there should be a strong concentration gradient between the spike (zone) and the rest of the mobile phase that contains none of the component. The diffusion resulting during its passage through the bed is called *longitudinal molecular diffusion*.

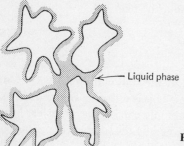

Liquid phase

Figure 4. Formation of pools of liquid phase between particles.

Longitudinal Molecular Diffusion

Sample molecules will diffuse from the region of high concentration (initial spike or center of the zone) into the region of lower concentration according to Fick's law. Expressing the diffusion in terms of the quarter-band width σ_{lmd}, we get

$$\sigma_{\text{lmd}} = (2\psi D_M t_M)^{\frac{1}{2}} \tag{20}$$

where D_M is the diffusion coefficient for the solute in the mobile phase and t_M is the time the sample spends in the mobile phase during its passage through the column of length L. This equation is just the Einstein relationship given in equation 6 except that an obstruction factor, ψ, has been added to account for the irregular flow in a packed bed. Note that t_M will be the same for all samples regardless of their partition coefficients since all samples spend the same amount of time in the mobile phase. This time is the time required to pass through a given length of bed at a constant linear velocity; that is, $t_M = L/\bar{u}$ (equation 20 in Chapter 2).

Eddy Diffusion

The other type of diffusion occurring in the mobile phase is called eddy diffusion or convective mixing. This is a classical concept that allows for the different paths that individual solute molecules will take in passing through a bed with irregular channels. Since different molecules take different paths, they will arrive at the end of the bed at different times, and the zone as a whole will be widened. The quarter-zone width can be expressed as

$$\sigma_{\text{ed}} = (2\lambda d_p L)^{\frac{1}{2}} \tag{21}$$

where d_p is the average diameter of the particles in the bed, L is the length of the bed, and λ is a factor that is determined by the packing characteristics of bed. The latter depends on the average size of particles, the range of actual sizes, the density or tightness of packing, and the like.

Mass Transfer

If we now consider a sample that is sorbed by the stationary phase ($K_p > 0$), the process of sorbing and desorbing occurs. This is diffusion from one phase to the other and back again, and it is commonly called mass transfer. Mass transfer in the mobile phase is the process whereby the sample (in the mobile phase) is brought to the surface of the stationary phase from the bulk mobile phase. Mass transfer in the stationary phase is analogous. A simple diagram will illustrate the mechanism of mass transfer between phases and why this process results in zone broadening.

Figure 5a shows a sample in a chromatographic bed. The upper zone represents the distribution of sample in the mobile phase, and the lower one in the stationary phase. If the partition coefficient is 2, twice as much sample will be dissolved in the stationary phase as in the mobile phase. As shown in Figure 5a, the sample is at equilibrium.

The situation an instant later is shown in Figure 5b, where the movement of the mobile phase has carried some sample with it down the bed and ahead of the zone mean, thus causing spreading. In order to restore the system to equilibrium, some of the sample in the leading edge of the mobile zone must sorb on the stationary phase, and some sample in the tailing edge of the stationary zone must desorb into the mobile phase. This mass transfer or

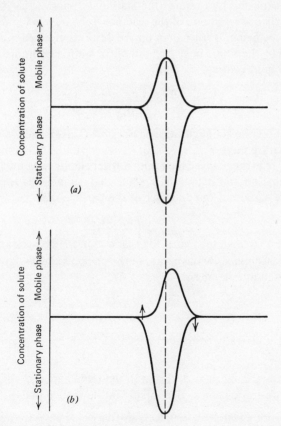

Figure 5. Illustration of zone spreading due to mass transfer ($K_p = 2.0$).

diffusion will result in a spreading of the zone, and we can express the extent of this action by the quarter-band width as before.

The simplest case to consider is that in which the stationary phase is a solid and the process is adsorption. Assuming that the adsorption and desorption rates are both first order and equal, Giddings (6) has derived the following general equation:

$$\sigma_{ad} = [2R_R (1 - R_R)u \ t_d \ L]^{1/2} \tag{22}$$

All symbols are the same as before, and t_d is the average time of desorption (or adsorption).

If the stationary phase is a liquid, the two mass-transfer rates are not equal, and separate equations must be considered for the diffusion-controlled mass transfer in the mobile phase and in the stationary (liquid) phase.* The equation for mass transfer in the stationary phase is very similar to the previous one for adsorption–desorption. The main difference arises from the different mechanisms. In a liquid (the stationary phase), the diffusion time depends directly on the square of the distance to be traveled and inversely on the diffusion coefficient, rather than on the desorption kinetics. The distance to be traveled by the solute in getting out of the stationary liquid is simply the depth of the liquid coating, d_f. Thus

$$\sigma_{mts} = \left[\frac{qR_R(1-R_R)ud_f^2 L}{D_S}\right]^{1/2} \tag{23}$$

where q is the so-called configuration factor and D_S is the diffusion coefficient in the stationary phase.

Mass transfer in the *mobile* phase does not depend on the retention ratio (or partition ratio), so the equation is somewhat simpler. However, it does depend on the square of the diameter of the particles d_p:

$$\sigma_{mtm} = \left(\frac{\omega d_p^2 uL}{D_M}\right)^{1/2} \tag{24}$$

ω is a factor determined by the nature of the packing and D_M is the diffusion coefficient in the mobile phase.

Summary

Four types of kinetically controlled processes in chromatography have been briefly presented:

1. Longitudinal molecular diffusion in the mobile phase
2. Eddy diffusion or convective mixing in the mobile phase

---

* It should be noted that the mass-transfer rate in the mobile phase is *not* the same as the mobile-phase velocity u, nor is it the solute velocity v. These are entirely different concepts.

3. Mass transfer in the stationary phase
4. Mass transfer in the mobile phase

In Chapter 8 they will be combined into one equation describing the extent of zone broadening for chromatography. It is called the van Deemter equation or the "rate" equation of chromatography.

ELECTRODIFFUSION

The only electrodiffusion technique we consider is zone electrophoresis. As already described, a voltage is imposed across a solution held on a solid support (e.g., paper). In such a field the rate of travel of an ion is expressed as its ionic mobility θ:

$$\theta = \frac{v}{E/L} \tag{25}$$

where v is the velocity of the ion and E/L is the voltage gradient in volts per centimeter.

If the ion can be assumed to be spherical and Stokes's law is applied, the ionic mobility can also be expressed in terms of other variables:

$$\theta = \frac{q^{\pm}}{6\pi r \eta} \tag{26}$$

where r is the radius of the ion, $q^{\pm}$ is its charge, and η is the viscosity of the solvent.

Another equation can be derived from equation 25 if (s/t) is substituted for the velocity and the equation is rearranged. (In this case s is the distance traveled in the time t.)

$$s = \frac{\theta E t}{L} \tag{27}$$

That is, the distance of migration is directly proportional to the ionic mobility, the voltage gradient, and the time.

Unfortunately this distance of migration will be influenced by other factors. Molecular diffusion as we have described it will also occur since a concentration gradient is also present. Furthermore, there is often some osmotic flow of solvent due to the voltage gradient as well as a chromatographic partitioning. Consequently it is not possible to describe zone electrophoresis with a single equation, and little use is made of this theoretical equation in practice. Further discussion of zone electrophoresis can be found in Chapter 16.

SLOW KINETIC PROCESSES

We have not and will not discuss the separations that are made possible by slow reaction rates. For example, in LLE there are cases in which one material is extracted virtually instantaneously while another is extracted slowly. By equilibrating the phases for only a short time, only the material with the fast rate is extracted, even though the K_p values may be the same. Precipitation is another common example in which widely differing rates permit separations of materials of nearly equal partition coefficients (solubility products).

REFERENCES

1. J. C. Giddings, *Dynamics of Chromatography*, Dekker, New York, 1965, pp. 227–230
2. F. Daniels and R. A. Alberty, *Physical Chemistry*, 3rd ed., Wiley-Interscience, New York, 1966, pp. 401–404.
3. B. L. Karger, in *Modern Practice of Liquid Chromatography*, J. J. Kirkland (ed.), Wiley-Interscience, New York, 1971, Chapter 1, p. 34.
4. S. Dal Nogare and R. S. Juvet, Jr., *Gas–Liquid Chromatography*, Interscience, New York, 1962, p. 135.
5. S. J. Hawkes, in *Recent Advances in Gas Chromatography*, I. I. Domsky and J. A. Perry (eds.), Dekker, New York, 1971, Chapter 2.
6. J. C. Giddings, in *Recent Advances in Gas Chromatography*, I. I. Domsky and J. A. Perry (eds.), Dekker, New York, 1971, p. 38.

PHYSICAL AND
MECHANICAL FORCES
IN SEPARATIONS

5

This chapter and the next present general discussions related to the forces between substances insofar as they are related to separation methods. The material is arranged to follow the classification system presented in Chapter 2, Table 5: namely mechanical, physical, and chemical. The first two types are included in this chapter, and chemical methods are in Chapter 6. In addition some discussion of the following separation methods is included in these chapters: formation of inclusion compounds, complexation and masking, precipitation, and ion exchange.

With the exception of the methods just mentioned, most of the discussion will be general. However, Table 1 presents a classification of methods according to type of interaction; it indicates where the general discussion can be applied. Those methods that are listed together have similarities that will allow information from one to be extended to the others. Thus a discussion

Table 1. Classification of Methods by Type of Interaction

Bulk Solution	Interface
Ionic:	Ionic:
Precipitation	Precipitation
Masking	Dialysis
Dialysis	Zone electrophoresis
Liquid–liquid extraction	
Nonionic:	Nonionic:
Gas–liquid chromatography	Gas–solid chromatography
Liquid–liquid chromatography	Liquid–solid chromatography
Liquid–liquid extraction	Inclusion compounds
Distillation	
Dialysis	

about ionic interfacial adsorptions on precipitates could be related to adsorption on a dialysis membrane as well as on the support used in zone electrophoresis.

PARTITIONING

According to the definitions in Chapter 2, the two types of partitioning are *ab*sorption (bulk phase) and *ad*sorption (interfacial). These two subclassifications are discussed separately after a general discussion about intermolecular forces.

Intermolecular and Interionic Forces

A discussion of intermolecular and interionic forces has to be so simplified as to lose some of its value; the subject is much too complex and our understanding is too inadequate. In the end we must rely heavily on empirical relationships. Still the knowledge we do have permits us to describe some phenomena and provides the basis for further investigations. Obviously this is an area that is still developing and changing.

What are some properties of molecules and ions that could be used to describe the forces between them? Some common ones are the following:

1. Ionization potential
2. Electron affinity
3. Electronegativity
4. Molar volume
5. Ionic radius
6. Ionic potential (charge-to-radius ratio)
7. Coordination number
8. Dipole moment
9. Dielectric constant
10. Polarizability
11. Boiling point and vapor pressure
12. Solubility
13. Activity coefficients
14. Solubility parameter of Hildebrand
15. Equilibrium constants

In most cases these parameters define a property of a given molecule or ion as it exists alone (neat) or perhaps in aqueous solution. For example, the high boiling point of water is commonly used as an illustration of the effect of intermolecular hydrogen bonding in drastically increasing the forces between water molecules—hence the boiling point. The boiling point is an example of a measure of the forces between like molecules, but in our study of separations we are more interested in forces between molecules of different substances. This fact makes our task that much more difficult since we know less about forces between unlike molecules and ions. As already noted, this often forces us to rely on empirical relationships.

Two qualitative terms not included in the previous list should be mentioned before proceeding. They are "polar" and "nonpolar." Both are so nonspecific that their use should be restricted. Whenever possible a more precise description should be given. This qualitative description has, however, produced equally nonspecific descriptions of intermolecular forces, such as "like dissolves like." This generalization has some utility, but often we can, and should, be more specific.

Ionic Forces. All of the forces to be studied are electrostatic in nature. For ions, the forces between ions of like charge are repulsive and between ions of different charge are attractive according to Coulomb's law:

$$E_{\text{Coulomb}} = \frac{q^+ q^-}{s^2} \tag{1}$$

where E_{Coulomb} is the energy of attraction, q is the charge (positive or negative), and s is the distance of separation. Compared to other energies, the coulombic attraction is a long-range force, as indicated by the dependence on s^{-2}. By comparison the repulsion energy is dependent on s^{-7} to s^{-12} and is only significant at very short distances. For our discussions we neglect these short-range repulsion energies.

van der Waals Forces. Three types of weak force have been identified, and together are called van der Waals forces. They can be named by the type of interaction or the man who first described them. Both systems are listed in Table 2.

Molecules that have a permanent dipole will obviously attract each other The attractive forces will cause the dipoles to line up with opposite poles attracting. Working against this alignment is the normal, random thermal motion, which keeps the attractive force small. The net energy of attraction has been found by Keesom to be

$$E_{\text{Keesom}} = -\frac{2}{3} \frac{\mu^4}{s^6} \frac{1}{k_B T} \tag{2}$$

Table 2. **Classification of Nonionic Intermolecular van der Waals Forces**

Interaction	Name	Investigator
Dipole–dipole	Orientation	Keesom (1912)
Dipole–induced dipole	Induction	Debye (1920)
Induced dipole–induced dipole	Dispersion	London (1930)

for molecules of the same type and

$$E_{\text{Keesom}} = -\frac{2}{3}\frac{\mu_A^2\mu_B^2}{s^6}\frac{1}{k_B T} \tag{3}$$

for molecules of different substances. In these equations μ is the dipole moment, k_B is Boltzmann's constant, and T is the absolute temperature. The negative sign indicates an energy of *attraction*; the dependence on s^{-6} indicates that this is a short-range energy.

A molecule with a permanent dipole can also *induce* a dipole into a non-polar molecule somewhat like a magnet attracting nonmagnetic iron. This induction of a dipole depends on the so-called polarizability ρ of the nonpolar molecule. Molecules that have large, easily deformed electronic clouds have large polarizabilities. The equation developed by Debye is

$$E_{\text{Debye}} = -\frac{\rho\mu^2}{s^6} \tag{4}$$

where μ is the dipole of the polar molecule and ρ is the polarizability of the nonpolar molecule.

The third force is the most difficult one to explain by analogy with a magnetic attraction or normal electrostatics. One example of it is the force of attraction that exists in the rare gases, which are nonpolar and have a symmetrical charge distribution. It is believed that at any particular instant this symmetry is somewhat distorted due to the motion (and position) of the electrons, so that there is a momentary polarity. This momentary dipole can induce a dipole in a neighboring atom or molecule in such a way as to produce a net attraction. As we have already seen, such inductions depend on the polarizability of the molecule (or atom). London has shown that

$$E_{\text{London}} = -\frac{3}{2}\frac{I\alpha^2}{s^6} \tag{5}$$

for molecules of the same type and

$$E_{\text{London}} = -\frac{3}{2}\frac{\alpha_A\alpha_B}{s^6}\left(\frac{I_A I_B}{I_A + I_B}\right) \tag{6}$$

for molecules of different substances. Here I is the ionization potential. London forces should be the only attractive forces between nonpolar molecules. For this reason it is relatively easy to measure London forces, but attractive forces between polar molecules are combinations of nondispersion forces and cannot be easily sorted out.

Table 3.   Relative Magnitude of Intermolecular Forces

Substance	Polarizability $\rho(\text{Å}^3)$	Dipole Moment μ (debyes)	Molecular Volume (ml/mole)	Energy (kcal/pair; $\text{Å}^6 \times 10^{23}$)		
				Keesom	Debye	London
He	0.21	0	26.0	0	0	3.6
Ar	1.63	0	23.5	0	0	165
Xe	4.0	0	27.2	0	0	650
H_2	0.81	0	24.7	0	0	27
CO	1.99	0.12	26.7	0.008	0.13	160
HCl	2.63	1.03	23.4	44	13	265
HBr	3.58	0.78	29.6	15	10	440
HI	5.4	0.38	35.7	0.84	4	880
NH_3	2.24	1.50	20.7	200	24	165
H_2O	1.48	1.84	18.0	450	24	110

Table 3 contains some typical values for these three energies for a few atoms and simple molecules. Beginning with the noble gases, which are non-polar and thus have no orientation and induction forces, we can compare the dispersion forces. As the polarizability increases, the dispersion force increases. Similarly the symmetrical molecule hydrogen has no permanent dipole moment and only dispersion interactions.

The other molecules listed in Table 3 have permanent dipoles and hence all three types of force. For the series of hydrogen halides, as the size and polarizability increase, the dipole moment decreases; hence dispersion forces increase, but orientation and induction forces decrease.

Hydrogen Bonding.   Hydrogen bonds are formed between molecules containing a hydrogen atom bonded to an electronegative atom like oxygen or nitrogen. Such is the case in alcohols, amines, and water. These molecules can both donate and receive a hydrogen atom to form hydrogen bonds. Other molecules—such as ethers, aldehydes, ketones, and esters—are only proton acceptors; proton donors like alcohols are necessary in order to form hydrogen bonds with them. Unlike Keesom forces, hydrogen bonds are directionally oriented, not random. Hydrogen-bonded molecules are bound together and are not free to move about independently as are those held by Keesom forces.

They are also strong enough (about 5 kcal/mole) to be considered weak "bonds"—that is, chemical reaction products. The distinction between a "force" and a "bond" is not sharp, but in our classification weak hydrogen bonds will be grouped with the other forces. The strength of these forces does often cause problems of irreversibility in separation methods.

Charge Transfer.   Finally, there is a group of specific interactions in which two molecules combine to form a charge-transfer complex like A^+B^- (*1*). This force is called *charge* transfer, but actually it is *electron* transfer. One of the first types used in chromatography was between Ag^+ ions and olefins (see, for example, reference *2*). If a π-electron system is involved, the complex is often called a π-complex. In some cases the interactions are so strong that they must be considered to be chemical reactions. In some of these cases the resulting stable complexes form the basis for extractions; others are colored and form the basis for colorimetric analysis. Consult the references at the end of the chapter for more information.

Absorption

When we apply our discussion of intermolecular forces to a consideration of bulk-phase properties (absorption), we are in effect discussing what is usually referred to as "theories of solutions." If a solution follows Raoult's law, it is called ideal, but few solutions are ideal. Other solutions follow the nonideal Henry's law over a small range of concentrations. Another theoretical model of solutions has been proposed by Hildebrand and is called the theory of "regular solutions" (*3*).

Regular Solutions.   The theory of regular solutions assumes that, when two solutions are mixed, the entropy of mixing is zero and the partial molar volumes remain unchanged. Any deviations from ideality are attributed to the enthalpy of mixing. Hence a regular solution is one in which the only nonideality is caused by dispersion interactions. The theoretical equation that can be used in dilute solutions to calculate the solute (A) activity coefficient γ_A is

$$\Delta\mathscr{H}_{\text{mix}} = \Delta\mathscr{G}_{\text{mix}} = \mathscr{R}T \ln \gamma_A = \phi_1^2 V_A(\delta_1 - \delta_A)^2 \tag{7}$$

where ϕ_1 is the volume fraction of the solvent, V_A is the molar volume of solute A, and δ_1 and δ_A are the so-called solubility parameters for the solvent and solute, respectively.

This new parameter, the solubility parameter, is a measure of the internal pressure of a substance, defined as

$$\delta = \left(\frac{\Delta\mathscr{E}_{\text{vap}}}{\bar{V}}\right)^{\frac{1}{2}} \tag{8}$$

where $\Delta\mathscr{E}_{\text{vap}}$ is the molar evaporation energy and $\bar{V}$ is the molal volume of the liquid. Clearly the solubility parameter is a measure of the forces between molecules; it is the square root of the energy of vaporization per milliliter. Some solubility parameters are listed in Table 4.

Table 4.   Some Solubility Parameters δ and Solvent Parameters ϵ^0

Compound	δ	ϵ^0 (Alumina)
n-Pentane	7.1	0.00
n-Hexane	7.3	0.01
Diethylether	7.4	0.38
Cyclohexane	8.2	0.04
n-Propyl chloride	8.5	0.30
Carbon tetrachloride	8.6	0.18
m-Xylene	8.8	—
Ethyl acetate	8.9	0.58
Benzene	9.2	0.32
Chloroform	9.2	0.40
Dibutylphthalate	9.3	—
Propyl nitrile	9.96	—
Acetone	10.0	0.56
Isopropanol	11.4	0.82
Acetonitrile	12.1	0.65
Methanol	14.5	0.95
Polyacrylonitrile	15.4	—

Chromatographers have tried to use the theory of regular solutions to obtain better methods of explaining and predicting chromatographic retention. Rohrschneider (4) has been able to draw some limited conclusions for GC, and his paper lists D. E. Martire's earlier contributions. Keller, Karger, and Snyder (5) have discussed the extensions of the original concept to include terms for polar and hydrogen-bonding interactions. These modifications have permitted the concept to be applied beyond the narrow confines of dispersion interactions (nonpolar molecules) and promise to place chromatographic theory on a more rational basis. By itself, however, the modified regular-solution concept is not capable of predicting chromatographic parameters with sufficient accuracy for it to be used for that purpose. For liquid–liquid chromatography, Martire and Locke (6) have suggested another fundamental approach that combines both thermal (energy) and athermal (size) factors. Their approach, like others, suffers from a lack of availability of such data as molar volumes, critical temperatures, and solubility parameters. For the present we must still rely on the empirical approach.

Adsorption

Because adsorption is defined as partitioning occurring at an interface or surface, it is most common to assume that one phase is a solid, and most of our discussion will bear on liquid–solid and gas–solid interfaces. In passing, however, we should note that adsorption has been found at liquid–liquid and gas–liquid interfaces. A recent paper by Liao and Martire (7) summarizes the chromatographic studies of this adsorption and provides a good bibliography

Adsorption on solid surfaces is at least as complex as absorption. At the molecular level we must consider such factors as the distance of separation of the active sites relative to the size of the sorbing molecules; the heterogeneous nature of the solid surface, varying from active (high energy) sites to less active ones; the arrangement (orientation) of the sorbed molecules on the solid; and the competition between solute and solvent molecules for the active sites.

Chromatography has stimulated much work in adsorption studies, of which Snyder's is the most comprehensive (8). He has attempted to combine all the possible interactions in liquid–solid chromatography into one generalized equation:

$$\log(K_p)_i = \log V_d + E_a(S^0 - A_s\epsilon^0) \tag{9}$$

where $(K_p)_i$ is the partition coefficient for solute i. The adsorbent parameters are V_a, its volume (volume of the stationary phase), and E_a, a surface-activity function; the solute parameters are S^0, the dimensionless free energy of adsorption, and A_s, the adsorbent area required by the solute; the solvent (mobile-phase) parameter is ϵ^0, as defined by this equation. A summary of what Snyder calls linear elution adsorption chromatography (LEAC) can be consulted for further details (9).

Snyder's Solvent Parameter ϵ^0. All of the new parameters in equation 9 are interrelated and should not be used independently, but Snyder has provided more data on alumina than any other adsorbent, so his ϵ^0 values on alumina are widely quoted. When ϵ^0 values are applied to other adsorbents, they should be multiplied by the relative E_a values suggested by Snyder:

$$\epsilon^0 \text{ (silica)} = 0.77\epsilon^0 \text{ (alumina)} \tag{10}$$

$$\epsilon^0 \text{ (magnesia)} = 0.58\epsilon^0 \text{ (alumina)} \tag{11}$$

$$\epsilon^0 \text{ (Florisil)} = 0.52\epsilon^0 \text{ (alumina)} \tag{12}$$

The larger the value of ϵ^0, the less strongly adsorbed is a given solute from that solvent. Some values for alumina are given in Table 4. Pentane has a value of zero by definition. Although the concept of ϵ^0 is quasi-empirical, it has been found to bear a relationship to the theoretical solubility parameter (10). The approximate relationship can be seen from the values in Table 4, but the original paper should be consulted for details (see also Chapter 14).

The decimal fractions in equations 10, 11, and 12 indicate the relative surface activities for these adsorbents, taking alumina as unity. On this basis, the most active adsorbent of the four is Florisil.

Adsorption Isotherms. Traditionally, adsorption isotherms have been used to describe the equilibrium condition between the concentration of a given solute which is sorbed and that which is not sorbed, at a given temperature. For many adsorption systems the isotherms are not linear, and thus they have been used to describe the different types of nonideal intermolecular attractions.

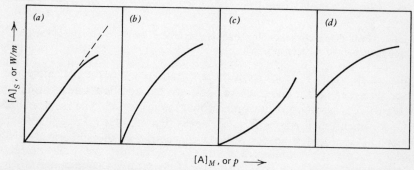

Figure 1. A comparison of four types of isotherm: (*a*) ideal; (*b*) Langmuir; (*c*) anti-Langmuir; (*d*) chemisorption.

The ideal adsorption isotherm is shown in Figure 1*a*. Along the ordinate is plotted the concentration of solute A that is sorbed ($[A]_S$ or W/m, where W is the weight of A sorbed and m is the weight of sorbent) and along the abscissa is plotted the concentration of solute that is unsorbed ($[A]_M$, the concentration of A in the mobile phase in chromatography, or p, the partial vapor pressure of the unsorbed solute). It shows the linear relationship between sorbed and unsorbed solute at a given temperature:

$$\frac{[A]_S}{[A]_M} = K_p = \frac{W/m}{p} \qquad (13)$$

The ideal case is seldom observed. Figures 1*b*, *c*, and *d* show three nonideal isotherms, although other more extensive classifications have also been suggested. The isotherm in Figure 1*b* follows the equation of Langmuir and represents a very common type:

$$\frac{W}{m} = \frac{k_1 p}{1 + k_2 p} \qquad (14)$$

where k_1 and k_2 are proportionality constants. This type of behavior is common in gas–solid and liquid–solid chromatography and can result from

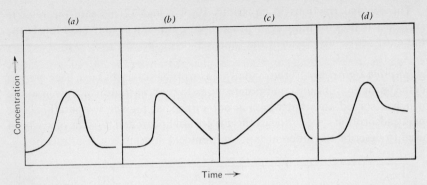

Figure 2.   Peak shapes corresponding to the isotherms in Figure 1: (*a*) ideal; (*b*) Langmuir; (*c*) anti–Langmuir; (*d*) chemisorption.

hydrogen bonding. It arises from those situations where the first molecules adsorbed cover up the most active sites, so that additional adsorption (at increased concentration of unsorbed solute) is decreased. The result of this behavior in chromatography is solute peaks that are tailed. Peak shapes are shown in Figure 2. However, at very low concentrations the isotherm is nearly linear.

A similar but simpler isotherm is the classical equation of Freundlich:

$$\frac{W}{m} = kp^{\frac{1}{x}} \tag{15}$$

where k is a proportionality constant and x is an integer greater than unity. The Freundlich isotherm differs from Langmuir's in that it does not approach a limiting value at high concentrations. In precipitation processes, adsorption on the newly formed precipitate (solid) is thought to follow the Freundlich isotherm.

Figure 1*c* shows the opposite type of adsorption (anti–Langmuir). In this case the first molecules adsorbed make it favorable for additional adsorption and sorption increases with concentration. In chromatography this results in leading peaks (Fig. 2). This is observed, for example, for flat molecules that stand on end when adsorbed, like phenol on alumina, as shown in Figure 3. It is also caused by overloading.

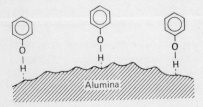

Figure 3.   Representation of the adsorption of phenol on alumina.

The fourth isotherm shown in Figure 1*d* represents chemisorption: a chemical reaction rather than a physical force. This can occur when the attractive forces are very strong. The main examples we have considered are strong hydrogen bonds and strong charge-transfer complexation. Other examples could include any type of chemical reaction of which acid–base is probably the most common. Many adsorbents have strongly acidic surfaces, which result in irreversible adsorption. Obviously chemisorption is undesirable in most cases, as can be seen from the poor peak shape obtained in chromatography (Fig. 2*d*).

MECHANICAL PROCESSES

Two types of mechanical separation processes are covered in this section: sieving and inclusion-compound formation. Both of them are methods that discriminate on the basis of molecular size and could be categorized as separations based on size.

Sieving

Two types of sieving can be distinguished. On the one hand the "sieve" allows molecules or ions below a certain size to pass through it while preventing the passage of larger ones. This is the case with dialysis, in which the semipermeable membrane serves as the sieve (see Chapter 17). On the other hand, the sieve can permit molecules or ions below a certain size to penetrate into its pores while the larger ones are prevented from entering and are swept away, as in chromatography. Later, the smaller molecules may diffuse out of the pores and reenter the stream, which is now free of larger particles. The principal example of this phenomenon is called molecular sieve chromatography.

For convenience, molecular sieve chromatography can be divided according to the state of the mobile phase—gas or liquid. In GC, molecular sieves are used primarily for the separation of fixed gases, a typical separation being hydrogen, oxygen, nitrogen, methane, carbon monoxide, ethane, carbon dioxide, and ethylene (*11*).

The sieves are natural zeolites or synthetic materials like alkali-metal aluminosilicates. When the latter are prepared, a lattice is formed that includes water of hydration. Driving off the water of hydration leaves pores or cages of a definite size determined by the lattice. Molecular sieves of varying pore size have been prepared (the common ones are listed in Table 5). Although the primary basis for separation with these sieves is size, adsorption does occur, especially for the small molecules that can enter the pores and

Table 5.   Some Molecular Sieves

Type	Pore Diameter (Å)
3A	3
4A	4
5A	5
10X	8
13X	9–10

experience the larger surface area. (More GC applications are given in Chapter 12, and a more general discussion of the analytical uses of molecular sieves can be found in reference *12*.)

In liquid chromatography the sieves being used are much more diverse, ranging from glass to synthetic polymers. In fact, there is no agreement on the name that should be used to describe the process, and at least 11 have been proposed (*13*). Two in particular represent the two somewhat separate areas of development: "gel filtration" and "gel permeation." The former name is associated with aqueous systems, and the latter with nonaqueous ones. The distinctions are further amplified in Chapter 14, but the processes are similar and should be classed under one name. "Molecular exclusion chromatography" is the one name that is becoming increasingly popular, and "molecular sieve chromatography" is usually intended to denote GC systems. A good summary is provided in reference *13*.

Exclusion chromatography is best suited for the separation of high-molecular-weight materials. The pore sizes of the sieves can be varied, and molecules ranging in molecular weight from several hundred to about 10^6 can be separated. However, the useful range for a given type of sieve is one to two orders of magnitude. Like the other molecular sieves, these also have active sites that result in some adsorption, especially for those molecules that enter the pores. Further discussion is deferred until Chapter 13.

Inclusion Compounds (*14*)

Inclusion compounds are stable crystalline solids that result from the combination of molecules in such a way that one (called the guest) is trapped inside the other (the host). Even the gas xenon forms white crystals when bubbled through carbon tetrachloride, chloroform, or acetone at $0°C$. A typical formula for one of these solids is $(CHCl_3)Xe_2(H_2O)_{17}$. In this and other examples no chemical reactions have taken place, but the guest is trapped in a "cage."

Two types of inclusion compound have been distinguished: *clathrates* are the type in which the guest molecule is completely surrounded in a cage by the host (*15, 16*); *adducts* are the type in which the guest molecule is trapped in a channel with open ends. A typical host that forms adducts is urea, but the term "adducts" is acknowledged to be confusing since it usually refers to the Diels–Alder type of adduct, which is different.

Inclusion compounds are relatively easy to form. The xenon compound whose preparation was described briefly is a typical clathrate. A simple preparation of a benzene clathrate for use in the undergraduate laboratory has been described (*17*); the clathrate is $Ni(CN)_2NH_3 \cdot C_6H_6$. The general procedure in forming adducts is to heat a mixture of the host and guest in a proper solvent and then allow it to cool, forming the cages. The solvents must be chosen carefully. They must dissolve both the host and the guest, but they must not act as guests and become trapped in the cages. For urea adducts methanol and mixtures of methanol and isooctane are good, but the water content is critical. A list of some guests that can be separated with urea is given in Table 6.

The bonding in inclusion compounds is definitely not chemical. The primary requirement is size: the proper guest in a given-size cage. This is the reason for classifying it as a sieving technique. However, there is no doubt that van der Waals forces are present and help stabilize the compounds, and in some cases hydrogen bonds are involved in forming the cages. Further information can be obtained from the references at the end of the chapter.

One of the main applications of urea adduction has been the separation of normal paraffins from petroleum stocks. Reference *18* describes a recent application to heavy distillates and provides a good summary of, and procedures for, urea adduction. Inclusion-compound formation has also been used for chromatographic separations. Examples are the GC separation based on clathrates formed with transition-metal complexes (*19*) and the TLC separation using urea (the host) as a part of the stationary phase (*20*).

Table 6. Some Guests Separated with Urea

Guest Type	Smallest Useful Member
Hydrocarbons	*n*-Hexane
Alcohols	*n*-Hexanol
Ketones	Acetone
Acids	Butyric acid
Halides	Octyl halide
Amines	Hexane diamine

REFERENCES

1. S. P. McGlynn, *Chem. Rev.* **58**, 1113 (1958).
2. B. L. Karger, *Anal. Chem.* **39** [8], 24A (1967).
3. J. H. Hildebrand and R. L. Scott, *Regular Solutions*, Prentice-Hall, Englewood Cliffs, N.J., 1962.
4. L. Rohrschneider, *J. Gas Chromatogr.* **6**, 5 (1968).
5. R. A. Keller, B. L. Karger, and L. R. Snyder, in *Gas Chromatography*, N. Stock and S. G. Perry (eds.), Institute of Petroleum, London, 1970.
6. D. E. Martire and D. C. Locke, *Anal. Chem.* **43**, 68 (1971).
7. H. Liao and D. E. Martire, *Anal. Chem.* **44**, 498 (1972).
8. L. R. Snyder, *Principles of Adsorption Chromatography*, Dekker, New York, 1968.
9. L. R. Snyder, in *Advances in Analytical Chemistry and Instrumentation*, Vol. 3, C. N. Reilley (ed.), Interscience, New York, 1964, p. 251.
10. R. A. Keller and L. R. Snyder, *J. Chromatogr. Sci.* **9**, 346 (1971).
11. See, for example, H. P. Burchfield and E. E. Storrs, *Biochemical Applications of Gas Chromatography*, Academic Press, New York, 1962, p. 185.
12. T. L. Thomas and R. L. Mays, in *Physical Methods in Chemical Analysis*, Vol. IV, W. G. Berl (ed.), Academic Press, New York, 1961, p. 45.
13. D. M. W. Anderson, F. C. M. Dea, and A. Hendrie, *Talanta* **18**, 365 (1971).
14. M. Baron, in *Physical Methods in Chemical Analysis*, Vol. IV, W. G. Berl (ed. Academic Press, New York, 1961, p. 223.
15. M. M. Hagan, *J. Chem. Educ.* **40**, 643 (1963).
16. V. M. Bhatnagar, *J. Chem. Educ.* **40**, 646 (1963).
17. G. D. Jacobs, *J. Chem. Educ.* **47**, 394 (1970).
18. J. R. Marquart, G. B. Dellow, and E. R. Freitas, *Anal. Chem.* **40**, 1633 (1968).
19. A. C. Bhattacharyya and A. Bhattacharjee, *Anal. Chem.* **41**, 2055 (1969).
20. V. M. Bhatnagar and A. Liberti, *J. Chromatogr.* **18**, 177 (1965).

SELECTED BIBLIOGRAPHY

General

Cadogan, D. F., and J. H. Purnell, *J. Chem. Soc.* **A9**, 2133 (1968).
Freeman, D. H., *Anal. Chem.* **44**, 117 (1972).
Jewell, D. M., and R. E. Snyder, *J. Chromatogr.* **38**, 351 (1968).
Maczek, A. O. S., and C. S. G. Phillips, *J. Chromatogr.* **29**, 15 (1967).
Meen, D. L., F. Morris, and J. H. Purnell, *J. Chromatogr. Sci.* **9**, 281 (1971).
Prout, C. K., and J. D. Wright, *Angew. Chemie Int. Ed.* **7**, 659 (1968).

Inclusion Compounds

Sebera, D. K., *Electronic Structure and Chemical Bonding*, Blaisdell, Waltham, Mass., 1964, p. 286.

Hagan, S. M., *Clathrate Inclusion Compounds*, Reinhold, New York, 1962.

Mandelcorn, L., *Chem. Rev.* **59**, 827 (1959).

Partition

Snyder, L. R., *Principles of Adsorption Chromatography*, Dekker, New York, 1968.

Moore, W. J., *Physical Chemistry*, 3rd ed., Prentice-Hall, Englewood Cliffs, N.J., 1962, pp. 713–716.

Adamson, A. W., *J. Chem. Educ.* **44**, 710 (1967).

CHEMICAL REACTIONS AND PROCESSES

6

Very few separation methods are based on chemical reactions; the most common one is precipitation. Consequently this chapter does not attempt a complete survey of chemical reaction types since this would be inappropriate. Even precipitation is not covered in detail because it is likely that this subject has already been covered by most students.

The chapter begins with a discussion of the chemical reactions of coordination complexes because they are important in a number of separation methods and also form the basis for a type of pseudoseparation called *masking*. Next, precipitation is discussed, with the main emphasis on the mechanisms of precipitation. Finally ion exchange is presented as a type of chemical reaction or chemisorption.

COORDINATION COMPLEXES

Many separations are made possible by a prior chemical reaction to form coordination complexes. The main examples are in liquid–liquid extraction, precipitation, ion-exchange chromatography, gas chromatography, as well as masking. Other analytical uses of coordination compounds include the formation of color or removal of color, the dissolving of precipitates, the changing of oxidation–reduction potentials, and serving as titrants. Coordination complexes are quite important in analytical chemistry, and many monographs have been written about them, including those listed at the end of this chapter. For that reason only a brief review of the fundamentals is presented here, followed by a short discussion of masking. When the individual separation methods are discussed, examples of the use of coordination complexes are given.

Basic Equilibria

The complexes to be discussed are formed between a central metal ion and other ions or molecules called ligands. As the metal ion is of primary interest, the separation methods under consideration are those intended to separate metal ions. Usually the solution being studied is aqueous.

Just as metal ions tend to be hydrated in aqueous solution, they attract other ions and molecules to form stable complexes called coordination complexes. This behavior is most prevalent for the transition metals. A typical equilibrium (neglecting other waters of hydration) can be written as

$$M^{n+} + xL \rightleftharpoons ML_x \tag{1}$$

where L is a ligand and the final charge on the complex ML_x depends on the charge (or lack of it) of the ligand. The species ML_x is called a mononuclear complex since only one metal ion is involved. More complicated situations exist in which more than one metal ion is involved (polynuclear complexes) or in which more than one type of ligand is involved (mixed-ligand complexes).

Even the equilibrium example in equation 1 is more complicated than shown if x is greater than unity. In those cases several equilibria can be written representing the individual stepwise addition of ligands:

$$M^{n+} + L \rightleftharpoons ML \tag{2}$$

$$ML + L \rightleftharpoons ML_2 \tag{3}$$

$$ML_2 + L \rightleftharpoons ML_3 \tag{4}$$

$$\cdots$$

$$ML_{(x-1)} + L \rightleftharpoons ML_x \tag{5}$$

The overall equilibrium represented by equation 1 is the sum of the individual steps, and the overall stability (equilibrium) constant is the product of the individual constants:

$$K_{stab} = \frac{[ML_x]}{[M][L]^x} = K_2 \times K_3 \times K_4 \times K_5 = K_1 \tag{6}$$

where K_1 is the constant for equation 1, K_2 is for equation 2, etc. In most cases the individual stability constants are of about the same order of magnitude, so many intermediate complexes can exist in addition to the final one ML_x.

Chelates

Since some ligands are capable of attaching at more than one point, the complexes that are formed contain one or more rings, usually of five members each, and are called chelates. One such ligand is ethylenediamine:

Ethylenediaminetetraacetic acid (EDTA) is a ligand with up to six groups that can attach to a single metal ion:

Consequently its reactions with metal ions go in one step and are simpler than the multistep reactions.

Table 1 lists some ligands and functional groups that can be part of chelating ligands. It can be seen from the list that many would be involved in other competing equilibria in aqueous solution—the most important being acid–base equilibria. For example, ammonia would be in equilibrium with the

Table 1. Some Ligands and Coordinating Functional Groups

Ion	Molecule	Functional Group	
		Ionic	Coordinate
F^-	NH_3	$-C{=}O$ with O^-	$-\overset{\displaystyle H}{\underset{\displaystyle H}{N{:}}}$
Cl^-	H_2O	$-O^-$	$\diagdown N{:}$
CN^-		$={N}^-$	$-NO_2$
SCN^-		$={N}-O^-$	$-NO$
OH^-		$={N}-O^-$ with O	$={N}-O-H$
			$\diagdown C{=}O$

ammonium ion. Thus a realistic study of coordination complexes can become quite involved since many of them are dependent on pH. One simplification suggested by Schwarzenback is the use of "conditional" equilibrium constants. Further information about conditional constants and more details about coordination equilibria can be found in most textbooks on ionic equilibria (see, for example, reference *1*).

Masking

Masking can sometimes be used in the analysis of a mixture of metal ions in lieu of a separation step. A complexing ligand is added to the solution so that the interfering metal ion is tied up or masked and prevented from causing an interference; that is, the normal chemical reaction of the interfering metal ion is prevented because the ion is converted (complexed) to another stable form (coordination complex). This process is also called *sequestering*. The ion is not completely removed because the equilibrium represented in equation 1 is still operative, but the concentration of the ion is significantly decreased, so that it does not interfere.

A list of masking agents taken from Meites (*2*) is given in Table 2. The papers by Cheng (*3*) and Hulicki (*4*) can also be consulted.

Where demasking is required, a number of procedures are possible. Addition of strong acid may be sufficient to shift the equilibrium, freeing the metal ion. Alternatively another stronger complexing agent can be added; or the ligand can be destroyed by chemical reaction or volatilized; or finally, the oxidation state of the metal ion can be changed (*3*).

PRECIPITATION

Precipitation is the classical method of separating metals. When it is combined with a measurement of the mass of the precipitate, the quantitative method of analysis is called *gravimetry*. A precipitating agent (or precipitant) is added to an aqueous solution containing the metal to be separated and reacts with it to form an insoluble solid according to the general reaction

$$a\mathrm{M}^{+x} + b\mathrm{X}^{-x} \rightleftharpoons M_a X_b \tag{7}$$

where M^{+x} is the metal ion and X^{-x} is the precipitant. Usually the system is allowed to come to equilibrium. The equilibrium constant, called the solubility product or ion product, K_{sp}, is expressed as the product of the ions remaining in solution:

$$K_{sp} = [\mathrm{M}]^a[\mathrm{X}]^b \tag{8}$$

Table 2. Some Masking Agents Used for Ions of Various Elements.[a]

Element (or Ion)	Masking Agents[b]
Ag	Br^-, Cl^-, CN^-, I^-, NH_3, SCN^-, thiourea
Al	OAc^-, BAL, Cit, $C_2O_4^{2-}$, EDTA, F^-, Mal, OH^-, salicylate, SSA, Tart, TEA, Tiron
As	BAL, OH^-, S_2^{2-}
Au	Br^-, CN^-, $S_2O_3^{2-}$
B	F^-, hydroxy acids
Ba	APCA, Cit, Tart
Be	F^-, Tart, SSA
Bi	APCA, BAL, Cit, Cl^-, dithizone, I^-, $Na_5P_3O_{10}$, Tart, TEA, thiourea
Br	Phenol (for Br_2)
Ca	APCA, Cit, F^-, polyphosphates, Tart
Cd	APCA, BAL, Cit, CN^-, dithizone, I^-, Mal, SCN^-, $S_2O_3^{2-}$, Tart
Ce	APCA, Cit, F^-, Tart, Tiron
CN^-	HCHO, Hg^{2+}
Co	APCA, BAL, Cit, CN^-, diethyldithiocarbamate, dimethylglyoxime, ethylenediamine, F^-, H_2O_2, Mal, $Na_5P_3O_{10}$, NH_3, NO_2^-, SCN^-, $S_2O_3^{2-}$, Tart
Cr	OAc^-, APCA, ascorbic acid, Cit, F^-, $NaOH + H_2O_2$, $Na_5P_3O_{10}$, SSA, Tart, TEA, Tiron
Cu	APCA, BAL, Cit, cobalticyanide, CN^-, I^-, NaH_2PO_2, NH_3, NO_2^-, S^{2-}, SCN^-, $S_2O_3^{2-}$, SSA, Tart, TGA
F^-	Al^{3+}, Be^{2+}, Fe^{3+}, H_3BO_3, Th^{4+}, Ti^{4+}, Zr^{4+}
Fe	APCA, ascorbic acid, BAL, $C_2O_4^{2-}$, citrate, CN^-, F^-, Mal, NH_3, $NH_2OH \cdot HCl$, oxine, 1,10-phenanthroline, PO_4^{-3}, $P_2O_7^{-4}$, S^{2-}, SCN^-, $S_2O_3^{2-}$, SSA, Tart, TEA, TGA, thiourea, Tiron
Ge	$C_2O_4^{2-}$, F^-
Hf	See Zr
Hg	Me_2CO, APCA, BAL, Cit, Cl^-, CN^-, I^-, SO_3^{2-}, Tart, TEA
Ir	Cit, SCN^-, Tart, thiourea
Mg	APCA, $C_2O_4^{2-}$, F^-, glycols, hexametaphosphate, OH^-, $P_2O_7^{-4}$, TEA

[a] From Meites (2).
[b] Abbreviations: OAc, acetate; Cit, citrate; EDTA, ethylenediaminetetraacetic acid; Mal, malonate; Tart, tartrate; APCA, aminopolycarboxylic acids; BAL, 2,3-dimercapto-1-propanol; SSA, sulfosalicylic acid; TEA, triethanolamine; TGA, thioglycolic acid.

Table 2 Some Masking Agents Used for Ions of Various Elements.[a] **(Continued)**

Element (or Ion)	Masking Agents
Mn	APCA, BAL, Cit, CN^-, $C_2O_4^{2-}$, F^-, $Na_5P_3O_{10}$, oxidants, $P_2O_7^{-4}$, SSA, Tart, TEA, Tiron
Mo	APCA, ascorbic acid, Cit, $C_2O_4^{2-}$, F^-, H_2O_2, $Na_5P_3O_{10}$, $NH_2OH \cdot HCl$, SCN^-, Tart, Tiron
Nb	$C_2O_4^{2-}$, F^-, H_2O_2, OH^-, Tart
NH_4^+	HCHO
Ni	APCA, Cit, CN^-, dimethylglyoxime, F^-, Mal, $Na_5P_3O_{10}$, NH_3, SCN^-, SSA, Tart
NO_2^-	Sulfanilic acid
Os	CN^-, SCN^-, thiourea
Pb	OAc^-, APCA, $(C_6H_5)_4AsCl$, BAL, Cit, I^-, $Na_5P_3O_{10}$, SO_4^{2-}, $S_2O_3^{2-}$, Tart
Pd	APCA, Cit, CN^-, I^-, NH_3, NO_2^-, SCN^-, $S_2O_3^{2-}$, Tart, TEA
Pt	APCA, Cit, CN^-, I^-, NH_3, NO_2^-, SCN^-, $S_2O_3^{2-}$, Tart
Rare earths	EDTA
Rh	Cit, Tart, thiourea
Ru	Thiourea
S	CN^-, S^{2-}, SO_3^{2-}
S^-	S
SO_3^{2-}	HCHO, Hg^{2+}
Sb	BAL, Cit, EDTA, I^-, OH^-, S^{2-}, S_2^{2-}, $S_2O_3^{2-}$, Tart
Sc	Tart
Se	S^{2-}, SO_3^{2-}
Sn	BAL, Cit, $C_2O_4^{2-}$, F^-, I^-, OH^-, Tart, TEA, TGA
Sr	APCA, Cit, SO_4^{2-}, Tart
Ta	Cit, F^-, H_2O_2, OH^-, Tart
Te	I^-
Th	APCA, Cit, F^-, SO_4^{2-}, 4-sulfobenzenearsonic acid, Tart, TEA
Ti	APCA, ascorbic acid, Cit, F^-, H_2O_2, $Na_5P_3O_{10}$, OH^-, SO_4^{2-}, SSA, Tart, TEA, Tiron
Tl	APCA, Cit, Cl^-, CN^-, $NH_2OH \cdot HCl$, Tart, TEA
U	Cit, CO_3^{2-}, $C_2O_4^{2-}$, F^-, H_2O_2, Tart
V	Ascorbic acid, CN^-, EDTA, F^-, H_2O_2, $NH_2OH \cdot HCl$, TEA, Tiron
W	F^-, H_2O_2, $Na_5P_3O_{10}$, $NH_2OH \cdot HCl$, SCN^-, Tart, Tiron
Zn	APCA, BAL, Cit, CN^-, dithizone, F^-, glycerol, glycol, $Na_5P_3O_{10}$, NH_3, OH^-, SCN^-, Tart
Zr	APCA, Cit, $C_2O_4^{2-}$, F^-, H_2O_2, PO_4^{-3}, $P_2O_7^{-4}$, SO_4^{2-}, Tart, TEA

Excess precipitant is often added to precipitate more completely the metal ion according to the "common ion effect." For effective separations K_{sp} should be 10^{-4} or less while other ions remain soluble. Tables of these equilibrium constants can be found in Meites (reference 2, Table I-9) and most textbooks of quantitative analysis.

For effective separations to be performed by precipitation the precipitant must be specific (i.e., precipitate only one ion) for the substance to be separated, or else the relative K_{sp} values for several ions must be sufficiently different to allow one (the least soluble one) to be precipitated while the rest remain in solution. Also additional complexing (masking) or pH regulation can be used to achieve the desired results. Unfortunately many newly formed precipitates do not obey the published equilibrium constants, and there is also considerable contamination due to coprecipitation and postprecipitation. As a result precipitation methods do not provide as good a separation as theory would indicate.

Supersaturation

To initiate precipitation the ion product (K_{sp}) has to be exceeded, sometimes by a large amount. To express this quantity we need to define supersaturation as the difference between the nonequilibrium ion concentration U and the equilibrium solubility S:

$$\text{Supersaturation} = U - S \qquad (9)$$

A better parameter is relative supersaturation:

$$\text{relative supersaturation} = \frac{U - S}{S} \qquad (10)$$

Some relative supersaturations needed to initiate precipitation are the following: AgCl, 1.7; SrSO$_4$, 10; BaSO$_4$, 32; and PbCrO$_4$, 45. There is an energy barrier to be overcome before crystallization can occur, just as there is an activation-energy barrier in chemical reactions, and supersaturation expresses the magnitude of this barrier.

Precipitates formed from solutions of high supersaturation are composed of small particles, often gelatinous, hard to handle, and badly contaminated, but this is not usually true of precipitates formed at low supersaturation. Small particles will have considerably more surface area per unit weight than large particles. Thus adsorption of foreign ions will be greater on small particles, and this accounts for some of the impurities. More information can be obtained by studying the mechanism of crystallization.

Mechanism of Crystal Formation

Crystallization is composed of two steps, nucleation and growth (5). Prior to nucleation there is continuous formation and dissolution of groups of ions called clusters. The size of these clusters is not known, but it is small (certainly less than 100 ions). The largest cluster that can exist before crystallization occurs is called the critical cluster. The addition of one more ion or molecule causes nucleation. Further addition of ions is called growth.

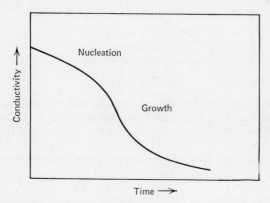

Figure 1.   An illustration of the precipitation process.

This process is shown in Figure 1, where a measure of the number of ions (conductivity) is plotted versus time. The two steps are indicated.

Nucleation predominates at high supersaturation, and growth predominates low supersaturation. Specifically,

$$\text{rate of nucleation} = k_n(U - S)^x \qquad (x > 1) \qquad (11)$$

$$\text{rate of growth} = k_g A(U - S) \qquad (12)$$

where A is the surface area of the particle. These two equations are plotted in Figure 2. The consequence of a large amount of nucleation (high supersaturation) is that many small nuclei are formed rather than a few larger crystals. Thus it is desirable to keep the supersaturation low.

Supersaturation can be kept low by using dilute solutions, although practical considerations represent a limit. Moreover, the reagents can be mixed with vigorous stirring to keep down local supersaturation. Homogeneous precipitation (to be discussed later) achieves the same result. Finally, the precipitation can be carried out at an elevated temperature, which will increase the solubility and decrease the adsorption of impurities.

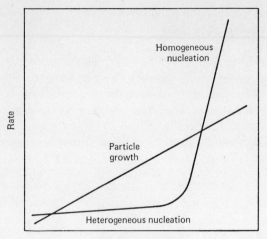

Figure 2. Relative rates of nucleation and growth. From *Fundamentals of Analytical Chemistry*, Second Edition by Douglas A. Skoog and Donald M. West. Copyright © 1963, 1969 by Holt, Rinehart and Winston, Inc. Reprinted by permission of Holt, Rinehart and Winston, Inc.

Impurities

"Coprecipitation" is the name given to the process that results in the contamination of precipitates during their formation and "postprecipitation" to the process that occurs after the precipitation is complete.

Coprecipitation includes inclusion, occlusion, and surface adsorption. Inclusion refers to foreign ions included in the crystal lattice. They can be isomorphic (mixed crystals) or nonisomorphic (solid solutions). Occlusion refers to foreign material physically entrained during the precipitation process, especially when colloidal particles are coagulated.

Surface adsorption follows the Paneth–Fajans rule: the ions that will be strongly adsorbed by a precipitate are those that form insoluble (or weakly dissociated) compounds with oppositely charged ions of the precipitate. This means that if a common ion is present (and one usually is in excess in precipitation), it will be preferentially adsorbed on the surface. Adsorption usually follows the Freundlich isotherm (Chapter 5).

Purification of contaminated precipitates can be effected by a reprecipitation, by digestion, and by washing during the filtration step.

Digestion is the process of allowing the precipitate to remain in contact with the mother liquor, usually at an elevated temperature. Under these conditions small particles dissolve and larger particles grow larger, thus decreasing surface impurities. Occluded impurities are released.

Washing should be carried out with a solution of an electrolyte to prevent peptization. The electrolyte must be volatile. Several washings (extractions) with small quantities is better than one washing with the same total amount of liquid (Chapter 10).

Homogeneous Precipitation

With some precipitation reactions it is possible to generate the precipitant in the solution. The result is called homogeneous precipitation as first described in 1937 by H. H. Willard and N. K. Tang. When the precipitant is generated this way, the supersaturation is kept very low, thus resulting in pure precipitates composed of large crystals.

Some reactions for generating precipitants are the following:

Sulfide:

$$CH_3-\overset{\overset{\displaystyle S}{\|}}{C}-NH_2 + H_2O \rightleftharpoons CH_3-\overset{\overset{\displaystyle O}{\|}}{C}-NH_2 + H_2S \tag{13}$$

Sulfate:

$$NH_2-SO_3H + H_2O \rightleftharpoons NH_4^+ + SO_4^{2-} + H^+ \tag{14}$$

Chromate

$$Cr^{3+} + BrO_3^- \rightleftharpoons CrO_4^{2-} + Br^- \tag{15}$$

8-Hydroxyquinoline:

Hydroxide:

$$NH_2-\overset{\overset{\displaystyle O}{\|}}{C}-NH_2 + H_2O \rightleftharpoons CO_2 + NH_3 \tag{17}$$

The hydrolysis of urea (equation 17) can be used not only to generate hydroxide ions to precipitate metal hydroxides but also as a general reagent to raise the pH, causing the precipitation of other insoluble metal salts (e.g., oxalates).

Because homogeneous precipitation offers many advantages, it has become the preferred method of separation by precipitation. Willard has coauthored a book on the subject (*6*), and several reviews have been published (*7*, *8*).

Examples

Most of the simple inorganic precipitates, such as the hydroxides and sulfides, are not very selective and hence of limited use. Table 3 lists some of the most common of these precipitates. Note that a few anions can be separated by precipitation.

Table 3.   Common Inorganic Precipitates

Separation of	Precipitate
Ag	AgCl
Ba	$BaSO_4$
Bi	Bi_2O_3
Fe	Fe_2O_3
K	K_2PtCl_6
Mg	$Mg_2P_2O_7$
Sr	$SrSO_4$
Cl	AgCl
SO_4	$BaSO_4$

Organic precipitants are much more selective and are usually preferred. Probably the best known is dimethylglyoxime (DMG), which can be specific for nickel. Flagg (*9*) has classified organic precipitants as (1) salt forming, including normal salts such as DMG and 8-hydroxyquinoline; and (2) adsorption type, which describes a small group of precipitants like tannin for which stoichiometric reactions cannot be written. He also includes a thorough discussion of about 20 of the most common organic precipitants.

A good summary is provided in Meites's handbook (*2*), which includes tables by element (Table 3-2), by precipitant (Table 3-1), for homogeneous precipitation (Table 3-3), and one emphasizing interferences (Table 3-6).

Evaluation

The advantages and disadvantages of precipitation separations are listed in Table 4.

Table 4.   Evaluation of Precipitation

Advantages	Disadvantages
1. Simple and inexpensive 2. Can handle large quantities; preparative 3. Highly accurate (parts per thousand) when combined with gravimetry	1. Not very efficient for most metals; fractional precipitation is usually not satisfactory; only one metal or one group of metals can be separated at one time 2. Not widely applicable; limited largely to metals 3. Large quantities required 4. Relatively slow and tedious

Crystallization

Crystallization of a substance from solution is not a chemical reaction, but it is included here because of the similarities between the crystallization process and precipitation. In precipitation, supersaturation is achieved by adding a precipitant, whereas, in crystallization it is achieved by cooling the solution or evaporating some of the solvent. The nucleation and growth processes that follow in each case are very similar.

Crystallization does not find much use as a method of separation in the analytical laboratory. Little apparatus is required, but controlled cooling is generally not available; centrifugation is often useful.

The principles of fractional crystallization are the same as those for distillation. In the presentation of phase diagrams in Chapter 3 this parallel was drawn and some diagrams were presented for liquid–solid systems. A complete discussion of crystallization can be found in the Weissberger series (10), including a list of common solvents used to crystallize organic compounds. The most important feature to note is that relative freezing points bear little relationship to boiling points. Thus a mixture that cannot be separated by fractional distillation might be capable of being separated by fractional crystallization. In general, the freezing point increases with increasing symmetry, as can be seen from this example of isomeric octanes: 2-methylheptane freezes at $-109°C$, n-octane at $-57°C$, and 2,2′,3,3′-tetramethylbutane at $+101°C$. The common isomer 2,2′,4-trimethylpentane freezes at $-107°C$, which is considerably less than would be expected on the basis of this generalization.

A related separation method, which can be considered to be the inverse of crystallization, is *zone melting*. The same phase diagrams and principles of crystallization apply. Zone melting is not included in this brief introduction (see the references at the end of Chapter 2). Both of these topics are covered in Karger, Snyder, and Horvath (*11*).

ION EXCHANGE

Most ion-exchange separations are carried out chromatographically in aqueous solutions, so ion-exchange chromatography is included in Chapter 14. Some of those separations, such as the separation of amino acids, are effected largely by ion–ion and ion–dipole forces like those discussed in Chapter 5, but others are effected by chemical reactions. The latter type are discussed here briefly and are called *chemisorption reactions*.

A typical ion-exchange reaction can be written as

$$B^+ + C^- + \underline{A^+R^-} \rightleftharpoons A^+ + C^- + \underline{B^+R^-} \tag{18}$$

where R^- represents a solid resin with an ionic site (anionic in this case). Since R^- is insoluble and represents the stationary phase in chromatography, it is the bed that is packed into the column; the other ions (B^+, C^-, A^+) flow past it and may be attracted to, and held by, it. In reaction 18 the resin has a negative charge and has exchanged a B^+ ion for its original A^+ ion and hence is called a cation-exchange resin. A similar reaction could be written for the opposite case of an anion-exchange resin. Commercially available resins are synthetic polymers and are described in Chapter 14.

Equation 18 can be considered to represent a *chemical* reaction, which would be called chemisorption. The sample is composed of $B^+ + C^-$ plus other ions, and B^+ is held by the resin while displacing A^+ from it. If the equilibrium is far to the right, which is to say that B^+ is more tightly held than A^+, B^+ will remain on the resin until displaced by another cation. Thus this is not an example of elution chromatography. It is a useful chemisorption-type ion-exchange reaction, a typical example of which is the softening of hard water. In this example a separation (of some ions from water) is achieved, but it is not chromatographic.

However, it is possible to achieve a chromatographic separation with a very similar system. The Li^+, Na^+, and K^+ ions can be separated on a cation resin in the H^+ form, where 0.7 *M* HCl is used as the mobile phase (*12*). In this case equation 18 could be written for the Na^+ equilibrium as

$$Na^+ + Cl^- + \underline{H^+R^-} \rightleftharpoons H^+ + Cl^- + \underline{Na^+R^-} \tag{19}$$

Since the resin is continually washed with HCl, the Na^+ ion, which is initially sorbed, is later desorbed according to the equilibrium conditions required by equation 19. The change in the nature of the mobile phase (and the discontinuous method of sampling) distinguishes this example from the preceding one. Unfortunately ion-exchange resins are not selective enough to make this type of separation practical over a wide range of metals.

It would be difficult to argue that equation 18 (and the example of water softening) represents chemisorption while equation 19 (and the example of elution chromatography) does not. On the other hand, one could describe ion-exchange chromatography on the basis of ion–ion and ion–dipole attractions. The distinction between these two classifications depends on one's definition of a bond and hence one's definition of what constitutes a chemical reaction. It was mentioned earlier that there is no clear distinction between a physical force and a bond, the difference being one of degree (note the case of the hydrogen bond). Hence ion exchange also falls into this intermediate category. The exact classification is not important for these discussions, but an understanding of the basis for the differing opinions is needed for perspective. The other feature that distinguishes equation 18 from equation 19 is the nature of the mobile phase; for equation 18 it is the ions on the left side of the equation, and for equation 19 it is the ions on the right side.

A large number of metal separations are carried out on ion-exchange resins by making use of coordination complexes. The most common one is the separation of Ni(II), Mn(II), Co(II), Cu(II), Fe(III), and Zn(II) (*13*) some form of which is included in many undergraduate chemistry laboratories. The sample containing these metals is put in 12 M HCl to form their chloro complexes, which are then put on an *anion*-exchange resin. A typical equilibrium would be

$$\text{Fe}^{3+} \overset{\text{Cl}^-}{\rightleftharpoons} \text{FeCl}^{2+} \overset{\text{Cl}^-}{\rightleftharpoons} \text{FeCl}_2^+ \overset{\text{Cl}^-}{\rightleftharpoons} \text{FeCl}_3 \overset{\text{Cl}^-}{\rightleftharpoons}$$
$$\overset{\text{Cl}^-}{\rightleftharpoons} \text{FeCl}_4^- \overset{\text{Cl}^-}{\rightleftharpoons} \text{FeCl}_5^{2-} \overset{\text{Cl}^-}{\rightleftharpoons} \text{FeCl}_6^{3-} \quad (20)$$

In 12 M HCl the highest complex, FeCl_6^{3-}, is formed, and the resulting anion is strongly held on the anion-exchange column. However, Ni(II) does not form an anionic chloro complex, so when the column is washed with 12 M HCl, Ni(II) is eluted. The other metals are held at the top of the column in the anionic form. Then the column is washed with successively less concentrated HCl (6 to 0.005 M), and the chloro complexes are destroyed, one by one, as the equilibria shift to the left (see equation 20). As soon as the charge on a given complex changes from negative to neutral (or positive), it is desorbed by the resin and flows through the column. A very good selective separation is achieved.

This type of separation depends on a chemical reaction, the formation of coordination complexes. It is one of the common forms of ion-exchange chromatography. However, it is not chromatography in the strictest sense since the materials eluted from the column (neutral species or cations) are not the same as those placed on the column initially (anions). Few people have made this distinction, but it needs to be made if an orderly classification is to be achieved.

REFERENCES

1. J. N. Butler, *Ionic Equilibrium, A Mathematical Approach*, Addison-Wesley, Reading, Mass., 1964, pp. 261–393.
2. L. Meites, Handbook of Analytical Chemistry, McGraw-Hill, New York, 1963.
3. K. L. Cheng, *Anal. Chem.* **33**, 783 (1961).
4. A. Hulicki, *Talanta* **9**, 549 (1962).
5. A. G. Walton, *Science* **148**, 601 (1965).
6. L. Gordon, M. Salutsky, and H. H. Willard, *Precipitation from Homogeneous Solution*, Wiley, New York, 1959.
7. F. H. Firshing, *Talanta*, **10**, 1169 (1963).
8. F. H. Firshing, *Adv. Anal. Chem. Instrum.* **4**, 1 (1965).
9. J. F. Flagg, *Organic Reagents*, Interscience, New York, 1948.
10. R. S. Tipson, in *Technique of Organic Chemistry*, 2nd ed., Vol. III, Part 1, A. Weissberger (ed.), Interscience, New York, 1956, pp. 395–562.
11. B. L. Karger, L. R. Snyder, and C. Horvath, *An Introduction to Separation Science* Wiley-Interscience, New York, 1973.
12. W. Rieman, *Record Chem. Progr.* (*Kresge-Hooker Sci. Lib.*), **15**, 85 (1954).
13. K. A. Kraus and G. E. Moore, *J. Am. Chem. Soc.* **75**, 1460 (1953).
14. D. A. Skoog and D. M. West, *Fundamentals of Analytical Chemistry*, 2nd ed., Holt, Rinehart and Winston, New York, 1969, p. 166.

SELECTED BIBLIOGRAPHY

Coordination Complexes and Masking

Bailer, J. C., Jr. (ed.), *Chemistry of the Coordination Compounds*, Reinhold, New York, 1956.

Basolo, F., and R. G. Pearson, *Mechanism of Inorganic Reactions*, 2nd ed., Wiley, New York, 1967.

Berg, E. W., *Physical and Chemical Methods of Separation*, McGraw-Hill, New York, 1963, Chapter 18.

Dean, J. A., *Chemical Separation Methods*, Van Nostrand–Reinhold, New York, 1969, Chapter 2.

Dwyer, F. P., and D. P. Mellor (eds.), *Chelating Agents and Metal Chelates*, Academic Press, New York, 1964.

Marcus, Y., and A. S. Kertes, *Ion Exchange and Solvent Extraction of Metal Complexes*, Wiley, New York, 1969.

Martell, A. E., and M. Calvin, *Chemistry of the Metal Chelate Compounds*, Prentice-Hall, Englewood Cliffs, N.J., 1952.

Perrin, D. D., *Organic Complexing Reagents*, Interscience, New York, 1964.

Perrin, D. D., *Masking and Demasking of Chemical Reactions*, Wiley, New York, 1970.

Ringbom, A., in *Treatise on Analytical Chemistry*, Part I, Vol. 1, Chapter 14, I. M. Kolthoff and P. J. Elving (eds.), Interscience, New York, 1959.

Ringbom, A., *Complexation in Analytical Chemistry*, Wiley-Interscience, New York, 1963.

Taube, H., *Chem. Rev.* **50**, 69 (1952).

Welcher, F. J., *Analytical Uses of EDTA*, Van Nostrand, New York, 1958.

Ion Exchange

Helfferich, F., *Ion Exchange*, McGraw-Hill, New York, 1962.

Kitchener, J. A., *Ion Exchange Resins*, Wiley, New York, 1957.

Kunin, R., *Ion Exchange Resins*, 2nd ed., Wiley, New York, 1958.

Marinsky, J. A., *Ion Exchange: A Series of Advances*, Dekker, New York, Vol. I, 1966; Vol. II, 1969.

Rieman, W., III, in *Physical Methods in Chemical Analysis*, Vol. IV, W. G. Berl (ed.), Academic Press, New York, 1961, pp. 133–222.

Rieman, W., and H. F. Walton, *Ion Exchange in Analytical Chemistry*, Pergamon Press, New York, 1970.

Samuelson, O., *Ion Exchange Separations in Analytical Chemistry*, Wiley, New York, 1963.

Precipitation

Gordon, L., M. Salutsky, and H. H. Willard, *Precipitation from Homogeneous Solution*, Wiley, New York, 1959.

Hermann, J. A. and J. F. Suttle in *Treatise on Analytical Chemistry*, Part I, Vol. 3, Chapter 32, I. M. Kolthoff and P. J. Elving (eds.), Interscience, New York, 1961.

Kolthoff, I. M., E. B. Sandell, E. J. Meehan, and S. Bruckenstein, *Quantitative Chemical Analysis*, 4th ed., Macmillan, New York, 1969.

Laitinen, H. A., *Chemical Analysis*, McGraw-Hill, New York, 1960.

Walton, A. G., *The Formation and Properties of Precipitates*, Wiley, New York, 1967.

DISTILLATION

7

The preceding chapters have dealt primarily with common principles of separation science, including brief discussions of the following techniques: inclusion-compound formation, masking, precipitation, crystallization, and ion exchange. This is the first chapter devoted entirely to a major separation technique. Note, however, that Chapters 8 and 9 distinguish it from the other major techniques covered in Chapters 10 to 17. Distillation precedes and is distinguished from the other techniques because it is a group II method whereas the others are group I methods (see Table 4 in Chapter 2). In normal operation the principles discussed in Chapters 8 and 9 are not applied to distillation.

In the analytical laboratory, distillation is most useful for screening liquid mixtures and for isolating large quantities of material (preparative methods). Very pure samples can be obtained, but it is uncommon to separate by distillation an entire complex mixture into its individual components. A more common use of distillation is the characterization of petroleum stocks into boiling-point ranges rather than separation into individual compounds.

Some simplifications of distillation theory are needed for an elementary treatment. For example, most discussion will be limited to two-component mixtures, even though the principles can be extended to more complex mixtures. Also a simple discussion of mass balance in the distillation column will require the assumption of total reflux operation or continuous sampling, or both, even though normal operation is performed at partial reflux with batch sampling.

SOME TERMS AND DEFINITIONS

As a liquid sample is heated and converted to vapor, some of it is condensed and runs back into the starting stillpot, and the rest of it is condensed and transferred to a collection vessel. The former is called the reflux, and the latter the distillate. The ratio of the quantity of reflux to distillate is called the reflux ratio,

$$\text{reflux ratio} = \frac{N_r}{N_d} \tag{1}$$

where N_r is the number of moles of reflux and N_d is the number of moles of distillate. During this process the rising vapor contacts the condensed liquid (reflux), providing the opportunity for equilibrium to be established between the two in accordance with Raoult's law (in the ideal case). This process is called *rectification*, and it is enhanced if a large surface area is provided for the vapor–liquid contact.

The rate of vapor passage through the column is called the *throughput*, T_r, and the actual amount (volume) of sample that is contained in the column (as vapor and reflux) at a given time is called the *holdup*. If the throughput becomes too high and/or the holdup is exceeded, the column can no longer function effectively and is said to be flooded.

APPARATUS

Several types of fractionating columns were shown in Chapter 2. The packing in the fractionating column is intended to increase the surface area, so it should be tightly packed for highest efficiency. However, the denser the packing, the greater the pressure drop across the column and the lower the throughput, so some compromise is usually necessary between efficiency and pressure drop. Moreover, dense columns with a high surface area retain more liquid sample and have a large holdup, which is undesirable; the holdup should be less than 10% of the total sample volume.

High-efficiency distillation apparatuses are more complex. Provision is made to heat and control the temperature of the fractionating column. A valve allows some of the distillate to be returned to the stillpot, thus regulating the reflux ratio (Figure 1). Connections are provided so that the apparatus can be evacuated for vacuum distillation. Finally, the fractionating column needs to be long to get a high efficiency, but since this results in an undesirably large holdup and pressure drop, some special types of column have been devised. Typical laboratory-scale stills have been described (*1*).

One of the special types of column has a spinning or rotating band. It is annular in design, with an inner rod that is rotated. A typical cross section in which the center rod has a spiral wound on it is shown in Figure 2. Note that this rod takes the place of packing used in other types of columns. Bands can be made of metal or Teflon. The distance between the rod and the outer (glass) wall is about 1 mm. The rod is spun at speeds of up to 7200 rpm, so that the spiral runs downward, throwing vapor against the wall where it contacts the descending liquid. Compared to an equivalent packed column, the spinning-band column has a low holdup, low pressure drop, and high efficiency. Some comparative values are given in Table 1.

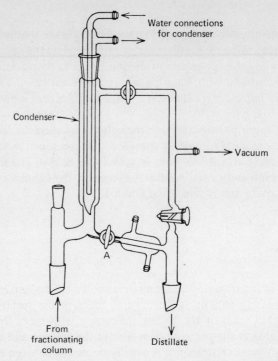

Figure 1. Distillation head for varying the reflux ratio. Stopcock A is for the adjustment of the reflux ratio.

Figure 2. Cross section of a spinning-band column.

Table 1. Comparison of Types of Fractionating Columns[a]

Type	Throughput (ml/min)	Holdup (ml/plate)	HETP (cm)
Vigreux	5–10	0.5–2	7–12
Glass helices	2–7	0.7–1	3–5
Metal helices	1–5	0.2–0.5	1–1.5
Concentric tube	0.5–2	0.02–0.03	0.5–1.0
Spinning band	3–5	0.01–0.03	0.5–3

[a] From Pecsok and Shields (2).

THEORY

The Theoretical Plate

It is possible to construct a fractionating column that has separate, individual steps or plates like that shown in Figure 3. Each plate corresponds to an evaporation–condensation step, which can be explained by reference to a temperature–composition phase diagram like the one shown in Figure 4.

If we begin with a mixture of A and B of composition 1 and heat it, boiling occurs at temperature T_1. The vapor produced has the composition represented by the intersection of the vapor-composition curve and the T_1 isotherm, namely, composition 2. When this vapor is cooled, a liquid of the same composition is obtained, as represented by a line dropped to the x-axis. Thus liquid 1 has been changed to liquid 2 by distillation. This process of going from liquid to vapor to liquid represents one plate. Continuing the process produces liquids in which the fraction of A is increasing. A two-plate column will produce distillate of composition 3, and a three-plate one of composition 4. Obviously a separation is being effected.

In a real bubble-cap column, each plate may not contain the liquid with the composition indicated in Figure 4 because each plate may not be as efficient as

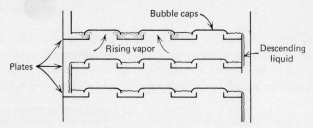

Figure 3.   Bubble-cap type of distillation column showing plates.

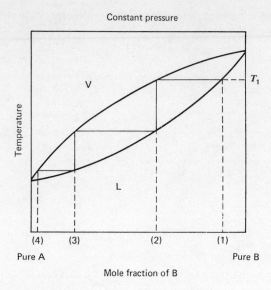

Figure 4. Temperature–composition phase diagram.

required by the theoretical figure. However, the image of a column with separate plates on which equilibrium can be established between liquid and vapor is useful in describing the concept of plates.

In actual practice it is more common to replot Figure 4 as shown in Figure 5 using the more volatile component (A in this case) as the basis. A 45-degree line is added, representing equal compositions of vapor and liquid. The procedure for counting the number of plates is the same as before, starting with the composition of the sample and proceeding to the desired final composition.

Since most fractionating columns are packed and do not have individual bubble-cap plates, it is customary to refer to "theoretical plates"; that is, a column that will produce liquid of composition 3 from starting material of composition 1 would be considered to have two theoretical plates. The symbol n will be used for the number of theoretical plates in distillation (and chromatography).*

Experimentally the number of theoretical plates in a column is usually measured by distilling a binary mixture and comparing the results with the phase diagram. A mixture of n-hexane and methylcyclohexane is often used, and details can be found in Weissberger (3). A student experiment with a

* It should be noted again that a given separation will require more chromatographic plates than distillation plates; that is, the number of theoretical plates required will differ in the two techniques.

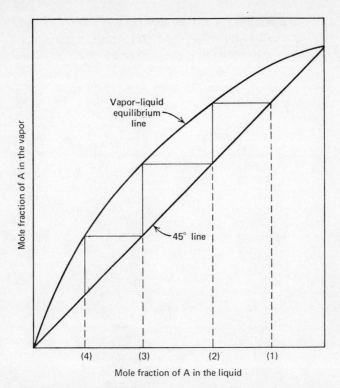

Figure 5. Composition diagram for determining the number of theoretical plates, A being the more volatile component of a binary mixture. Three theoretical plates are necessary to go from composition 1 to composition 4.

mixture of benzene and toluene has been described (4) as well as a separate experiment for determining the benzene–toluene phase diagram (5).

Another measure of column efficiency is the length of a theoretical plate or (since a distillation column is normally vertical) the *height equivalent to a theoretical plate* (HETP or just H). If the length (height) of the column is L,

$$H = \frac{L}{n} \tag{2}$$

The H-value has the advantage of being independent of L, unlike n. Some values are included in Table 1.

Theoretically, H can be expressed in terms of operating conditions as

$$H = \frac{11}{48} \frac{T_r s^2}{D} + \frac{D}{T_r} \tag{3}$$

where s is the radius of an open column or the average distance between the liquid and vapor phases, T_r is the throughput, and D is the diffusion co-efficient. Since a small H-value represents the greatest efficiency, T_r and s should be small in a good column; the second term, D/T_r, is often negligible.

Relative Volatility and the Fenske Equation

In discussions of distillation theory it is common to use the concept of volatility. It can be defined as the ratio of the mole fraction of component A in the vapor, $(X_A)_{vapor}$, to the mole fraction of A in the liquid, $(X_A)_{liq}$

$$\text{volatility} = \frac{(X_A)_{vap}}{(X_A)_{liq}} \tag{4}$$

or, since the partial vapor pressure p is another way of expressing vapor concentration,

$$\text{volatility} = \frac{p_A}{(X_A)_{liq}} \tag{5}$$

Using either definition 4 or 5, we can define a relative volatility for the binary mixture of A and B as

$$\text{relative volatility} = \frac{\text{volatility of A}}{\text{volatility of B}} \tag{6}$$

$$= \frac{(X_A)_{vap}(X_B)_{liq}}{(X_A)_{liq}(X_B)_{vap}} \tag{7}$$

$$= \frac{p_A}{p_B} \frac{(X_B)_{liq}}{(X_A)_{liq}} \tag{8}$$

According to Raoult's law,

$$p_A = (X_A)_{liq}\, p_0^A \tag{9}$$

where p_A^0 is the vapor pressure of A, so that equation 8 becomes

$$\text{relative volatility} = \frac{p_A^0}{p_B^0} = \alpha \tag{10}$$

which is the definition of the separation quotient α introduced in Chapter 2. The concept of relative volatility in distillation is the same as the broader concept of separation quotient.

Further, equation 7 can be rearranged to

$$\frac{(X_A)_{vap}}{(X_B)_{vap}} = \alpha \frac{(X_A)_{liq}}{(X_B)_{li}} \tag{11}$$

Since for a binary mixture

$$(X_A)_{vap} + (X_B)_{vap} = 1 \tag{12}$$

and

$$(X_A)_{liq} + (X_B)_{liq} = 1 \tag{13}$$

then

$$\frac{(X_A)_{vap}}{1 - (X_A)_{vap}} = \alpha \frac{(X_A)_{liq}}{(1 - X_A)_{liq}} \tag{14}$$

We have arbitrarily selected A as the more volatile component, so α will always be greater than unity. Thus, in equation 14, α can be considered to be an enrichment factor. It expresses the change in sample composition in going from liquid to vapor: the vapor is enriched in the more volatile component A.

For a fractional distillation with a column of n theoretical plates, it has been shown that

$$\frac{(X_A)_{vap}}{1 - (X_A)_{vap}} = \alpha^{n+1} \frac{(X_A)_{liq}}{1 - (X_A)_{liq}} \tag{15}$$

which is known as the Fenske equation. The exponent is $(n + 1)$ rather than n since one theoretical plate is obtained between the stillpot and the first plate in the fractionating column.

Distillation Under Partial Reflux

The methods of finding the number of theoretical plates and the equations in the last section all assume that the distillation is operated under total reflux. Obviously this is impractical since no benefits can be obtained if no distillate is produced. Still, it has been useful from the standpoint of theory.

There are two common methods for treating distillation under partial reflux. The first is the graphical procedure of McCabe and Thiele based on the "operating line equation" of Lewis. Strictly speaking it is applicable only to continuous distillation, in which the composition at each point in the fractionating column may be regarded as constant, but it also finds use in determining n for batch processes. Even though it was not derived for this use, the procedure is better than the one for total reflux that we have used (Fig. 5).

The second method is one that *does* apply to batch distillation. It is the Rayleigh equation, which is used to determine the composition of the liquid in the stillpot as the distillation proceeds (rather than the number of theoretical plates.) Both of these topics will be treated briefly.

McCabe–Thiele Plot. This method is based on the operating line equation of Lewis, which was derived under the assumption that the column is operated adiabatically, that the heat of mixing in the vapor and liquid phase is negligible, and that the substances involved have similar thermal properties. It is based on the simple relationship that the total number of moles of material leaving a given section of the system (e.g., a plate) must equal the total number of moles entering that section per unit time (under continuous distillation):

$$N_v = N_r + N_d \tag{16}$$

where N is the number of moles per unit time and the subscripts v, r, and d stand for vapor, reflux, and distillate, respectively.

The Lewis equation is applicable anywhere in the system as an expression of the mass balance. Take, for example, the mass balance between plate n and plate $(n-1)$:

$$N_v(X_A)_{\mathrm{vap},n-1} = N_r(X_A)_{r,\mathrm{liq},n} + N_d(X_A)_{d,\mathrm{liq}} \tag{17}$$

Rearranging, the operating-line equation is commonly written as

$$(X_A)_{\mathrm{vap},n-1} = \frac{N_r}{N_v}(X_A)_{r\mathrm{liq},n} + \frac{N_d}{N_v}(X_A)_{d\mathrm{liq}} \tag{18}$$

In the McCabe–Thiele procedure this line is added to Figure 5 as shown in Figure 6. The two points used to draw the line are the intersection of the 45-degree line with $(X_A)_{d\,\mathrm{liq}}$ and the intercept on the y-axis for $(N_d/N_v)(X_A)_{d\mathrm{liq}}$. It is interesting to observe that for total reflux operation, $N_d = 0$ and $N_r/N_v = 1$, so equation 14 reduces to

$$(X_A)_{\mathrm{vap},n-1} = (X_A)_{r\mathrm{liq},n} \tag{19}$$

which corresponds to the 45-degree line in Figures 5 and 6.

The number of theoretical plates is obtained from Figure 6 by the same stepping-off procedure we used before.

Rayleigh Equation. In normal laboratory batch operation, distillate is removed from the process; the composition of the phases is therefore continually changing. The variation in the composition of a binary mixture in the stillpot can be calculated with the Rayleigh equation:

$$\ln \frac{(N_A)_{\mathrm{still},1}}{(N_A)_{\mathrm{still},2}} = \int_{x=2}^{1} \frac{d(X_A)_{\mathrm{still},x}}{(X_A)_d - (X_A)_{\mathrm{still},x}} \tag{20}$$

where N_A is the number of moles of A in the still, X_A is the mole fraction of A, and 1 and 2 are the limits representing the time interval during which distillation has proceeded. This equation can be combined with the Fenske equation and used to examine the effect of variables on batch fractional distillation.

A typical plot showing the effect of reflux ratio is given in Figure 7 (6). Clearly the separation improves as the reflux ratio increases, but this in turn

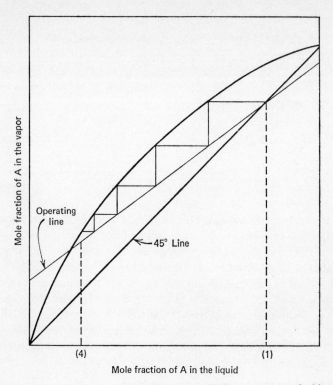

Figure 6. McCabe–Thiele plot, A being the more volatile component of a binary mixture. About 4.5 theoretical plates are necessary to go from composition 1 to composition 4.

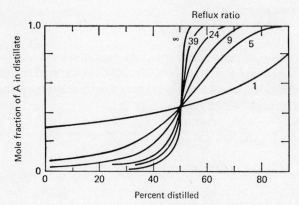

Figure 7. Effect of some variables on the distillation efficiency of an equimolar binary mixture. $n = 11$; $\alpha = 1.5$; $\alpha^n = 86.5$. *From Treatise on Analytical Chemistry*, Vol. 2, I. M. Kolthoff and P. J. Elving (eds.). Copyright by John Wiley & Sons.

Table 2. **Effect of α on the Required Number of Plates**

Parameter	Example			
	1	2	3	4
α	1.5	2	1.1	1.05
n	11	6.5	47	91
α^n	86.5	87.3	88.2	84.8

decreases the amount of distillate collected and increases the time required. A comparison of the other variables n, α, and α^n is shown in Table 2. In this table α is varied, and the number of theoretical plates required to keep the enrichment factor (α^n) constant is calculated for each case. The first example is the one shown in Figure 7. When α is raised to 2, the required number of plates is down to 6.5, but when α is decreased to 1.1 and then to 1.05, n increases to 47 and 91, respectively. Calculations like these are useful in describing distillation and in selecting the proper conditions. Recall from Chapter 3 that α can be estimated from boiling points according to the following equation:

$$\log \alpha = 8.9 \frac{T_B - T_A}{T_A + T_B} \tag{21}$$

PRACTICAL APPLICATIONS

In the average analytical laboratory one does not often have a wide range of distillation columns to choose from, and often the choice is determined by the size of the sample. Also, of course, most samples are much more complex than the binary mixtures just discussed.

However, the following general rules for a binary mixture can be drawn from the theory. On the basis of equation 21, Rose (6) has shown that

$$\text{optimum reflux ratio} = n \tag{22}$$

$$n = \frac{2.85}{\log \alpha} \tag{23}$$

$$= \frac{T_A + T_B}{3(T_B - T_A)} \tag{24}$$

Hence, beginning with the boiling points of the two components, follow this procedure:

1. Calculate α from the boiling points or vapor pressures.
2. Estimate n and the reflux ratio from equation 23. The reflux ratio is commonly measured by counting the number of drops of distillate and reflux. It is adjusted with a head like that shown in Figure 1.
3. Remember that separations are better when there is a larger proportion of the more volatile component, and vice versa.
4. Select the apparatus according to the quantity of sample available, the number of plates required, and with a view to keeping holdup, throughput, and packing spacing (r) small.

SPECIAL TECHNIQUES

Azeotropic Distillation

In Chapter 3 it was indicated that many binary mixtures are not ideal and go through a maximum or minimum in their temperature–composition curves. These maxima or minima are called *azeotropes*. When a minimum is present, it can prevent a distillation from being successful since the stepping-off procedure we have followed yields the azeotropic mixture rather than pure A. This is shown in Figure 8. Mixtures of composition 1 or 2 will ultimately yield

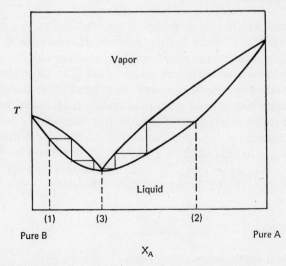

Figure 8. Phase diagram for a two-component system with a minimum boiling azeotrope.

a distillate with the composition of the azeotrope, 3. Separation into pure A and B is not possible.

However, this phenomenon can also be used to advantage. The classic example is the separation of ethanol and water, which results in a mixture of 95% ethanol with a boiling point of 78°C due to azeotrope formation. To get rid of the last 5% of water, a third component, called an *entrainer* (benzene in this case), is added. An azeotrope with a lower boiling point (65°C) is formed; it is composed of 74% benzene, 18.5% ethanol, and 7.5% water. With excess benzene, all of the water can be removed. Then the excess benzene is distilled off (actually another azeotrope of 67.6% benzene) at 68°C.

In the petroleum industry entrainers are added to facilitate group separations (like aromatics from paraffins and cycloparaffins) by forming minimum boiling azeotropes. The requirements of an entrainer are as follows:

1. A boiling point about 30°C less than that of the sample
2. Solubility in water and not much solubility in the hydrocarbon (at room temperature), so that the excess can be easily extracted away with water
3. Solubility in the hydrocarbon at the distillation temperature
4. Purity and low cost
5. Lack of reaction with the sample

Some typical entrainers are acetone, methanol, ethanol, acetic acid, and ethylene glycol ethers.

Extractive Distillation

Extractive distillation is similar to azeotropic distillation. An entrainer is added to change the α of the boiling mixture. However, in this case the entrainer is nonvolatile and has a preferential interaction with one component of the mixture. (This situation is very similar to GLC, in which the stationary phase exerts a selective influence on each component of the sample and thus permits a separation on a basis other than boiling point alone. In fact GLC has been used to select entrainers for extractive distillation.) Normally the entrainer is introduced at the top of the column in a continuous distillation, so it must be possible to separate it from the other components in the still and pump it to the top of the column. This is an obvious disadvantage. Extractive distillation is not much used in the analytical laboratory.

Vacuum Distillation

By reducing the pressure below atmospheric the boiling points of samples can be reduced. This is useful in those cases where decomposition could occur at the normal boiling points or where the high temperatures are difficult to achieve. An added benefit in some cases is that the relative volatility is also increased at a reduced pressure. This can be seen by referring to Figure 1 in Chapter 3. The slope of this plot is a function of the heat of vaporization, $\Delta \mathscr{H}$. If two components of a sample have different heats of vaporization, their plots will not be parallel and usually will converge at the higher temperature and pressure.

However, vacuum distillation has its disadvantages, and Rose (6) does not recommend it. The throughput is decreased, and columns tend to flood. In general, the variables are more difficult to regulate. At very low pressures ($<10^{-3}$ torr) molecular distillation takes over, and this is a different phenomenon.

Steam Distillation

It is possible to distill the components of an immiscible two-phase system at a temperature lower than that of their normal boiling points. Use is made of this phenomenon to distill relatively nonvolatile samples or samples that decompose at high temperatures. In this respect it is often used for the same reasons as vacuum distillation. Since water is usually used as the second immiscible phase, the process is called *steam distillation*. An external source of steam is attached to the stillpot so that it is always present in excess amounts. Distillation is continued until the desired component of the sample is exhausted.

EVALUATION

Fractional distillation is used mainly for screening liquid samples and for preparative work. Even a relatively crude distillation indicates the boiling

Table 3. Evaluation of Fractional Distillation

Advantages	Disadvantages
1. Good for preparing large quantities 2. Good for general screening of liquids 3. Relatively inexpensive and simple	1. Not very selective. Limited to separations based on vapor pressure. Not amenable to separation of all components 2. Undesirable azeotropes often formed 3. Some compounds are thermally labile 4. Slow 5. Large quantities of sample required 6. Not readily adapted to quantitative analysis

range of the sample and its complexity, but for precise multicomponent analysis it cannot compete with chromatography. Large quantities of very pure (>99.9%) chemicals can be prepared, especially with the spinning-band-type columns. The advantages and disadvantages are summarized in Table 3.

SUBLIMATION AND ZONE REFINING

The theory in this chapter is applied to liquid–vapor equilibria. Similar discussions apply to solid–liquid and solid–vapor equilibria, which form the basis of the separation techniques of zone refining and sublimation, respectively. Though these techniques are not included in this book, the similarities should be noted.

REFERENCES

1. J. C. Winters and R. A. Dinerstein, *Anal. Chem.* **27**, 546 (1955); A. G. Nerheim and R. A. Dinerstein, *Anal. Chem.* **28**, 1029 (1956); A. G. Nerheim, *Anal. Chem.* **29**, 1546 (1957).
2. R. L. Pecsok and L. D. Shields, *Modern Methods of Chemical Analysis*, Wiley, New York, 1968, p. 32.
3. A. Weissberger (ed.), *Technique of Organic Chemistry*, Vol. IV, Interscience, New York, 1951, p. 32.
4. A. Ault, *J. Chem. Educ.* **41**, 432 (1964).
5. N. J. Molski and H. A. Swain, Jr., *J. Chem. Educ.* **45**, 48 (1968).
6. R. Rose, in *Treatise on Analytical Chemistry*, Part I, Vol. 2, I. M. Kolthoff and P. J. Elving (eds.), Wiley–Interscience, New York, 1961, pp. 1264–1270.

SELECTED BIBLIOGRAPHY

Berg, E. W., *Physical and Chemical Methods of Separation*, McGraw-Hill, New York, 1963 Chapter 2.

Carney, T. P., *Laboratory Fractional Distillation*, Macmillan, New York, 1949.

Laitinen, H. A., *Chemical Analysis*, McGraw-Hill, New York, 1960.

Podbielniak, W. J., and S. T. Preston, in *Physical Methods in Chemical Analysis*, Vol. III, W. G. Berl (ed.), Academic Press, New York, 1956.

Rose, A., in *Treatise on Analytical Chemistry*, Part 1, Vol. 2, I. M. Kolthoff and P. J. Elving (eds.), Wiley–Interscience, New York, 1961.

Rose, A., and E. Rose, in *Technique of Organic Chemistry*, Vol. IV, A. Weissberger (ed.), Interscience, New York, 1951.

Turkal, P. J., *Am. Lab.* **3**[2], 50 (February 1971).

Extractive and Azeotropic Distillation, No. 115 in *Advances in Chemistry Series*, American Chemical Society, Washington, D.C., 1972.

ZONE BROADENING
IN MULTISTEP PROCESSES

8

Chapters 8 and 9 summarize the principles concerning the mechanics of separation methods that are performed in a multistep (or continuous) countercurrent fashion. These methods include those classified in Chapter 2 as group I methods—those in which the second phase is *added*. When they are carried out in a multistep countercurrent fashion, the results of the separation are usually presented as a series of peaks or bands with a symmetrical Gaussian shape. These methods are treated individually in Chapters 10 to 17.

The problem of separation can be reduced to a consideration of only two components. As each component passes through the separation system, it becomes diluted with the mobile phase, and the concentration profile (or zone) is broadened. The extent of this zone broadening, the shape of the zone, and the factors that influence them are covered in this chapter. At the same time, each component passes through the system at a given rate (which is a fraction of the mobile phase velocity) depending upon its interaction with the stationary phase. If the two components have different rates of travel, their zone centers will be separated. Consequently, the criterion for deciding if a separation has been achieved is a comparison of the distance of separation of the zone centers with the extent of broadening of each of the two zones. This comparison is the subject of Chapter 9.

Two approaches are taken to the subject of zone broadening. The first assumes individual, discontinuous steps, in which the system comes to equilibrium in each step. Of the four separation methods to be discussed, only countercurrent liquid–liquid extraction (LLE) is carried out this way, although this approach has been applied to continuous methods, notably chromatography. It is sometimes referred to as the "plate" model. The other approach concerns continuous countercurrent methods, in which the system does not come to equilibrium. It is sometimes called the "rate" model, and it has been applied to chromatography with considerable success.

To a first approximation; both of these models come to the same conclusions: zone broadening is proportional to the square root of the "length" of the system (number of steps, length of column, etc.) and the shape of the zone is Gaussian if the system is sufficiently long.

EQUILIBRIUM APPROACH (AS IN LLE)

Probably the best way to gain an understanding of zone broadening in countercurrent LLE is to consider a simple example. For this purpose we shall use a *single* substance rather than a mixture since we want to study zone broadening per se.

For our "system" we shall use a series of common separatory funnels and two immiscible solvents that will be referred to as extractant and raffinate. For simplicity, the volumes of the two solvents will be equal in each funnel, for example, 50 ml of each:

$$V_{\text{raff}} = V_{\text{extr}} = 50 \text{ ml} \tag{1}$$

The sample will be called A, and the sample size will be taken as 100 mg. We shall assume that sample A is preferentially soluble in the raffinate and has a partition coefficient in this system of 2.0 (at some constant but unspecified temperature). Then

$$(K_p)_\text{A} = 2.0 = \frac{[A]_{\text{raff}}}{[A]_{\text{extr}}} = \frac{(W_\text{A})_{\text{raff}}}{(W_\text{A})_{\text{extr}}} \times \frac{50}{50} \tag{2}$$

where W_A signifies the weight of A in the respective phases.

The countercurrent process is depicted in Figure 1 as a series of five separatory funnels. In the first step (zero transfer) 50 ml each of extractant and raffinate are placed in funnel 1. A sample of 100 mg of A is added to the raffinate (no further sample will be added to the system). Thus all of the sample is initially in the first funnel. Funnel 1 is shaken and allowed to come to equilibrium, at which time, according to the partition coefficient,

$$(K_p)_\text{A} = 2.0 = \frac{66.7}{33.3} \times \frac{50}{50} = \frac{66.7}{33.3} \tag{3}$$

In other words, 33.3% of the sample is extracted and 66.7% remains in the raffinate. These quantities are rounded off and entered on the respective portions of Figure 1.

In the first transfer, the extractant (33.3 mg of A in 50 ml) is moved to the second funnel. A 50-ml quantity of fresh extractant is added to the first funnel (which still contains the raffinate and 66.7 mg of A) and 50 ml of fresh raffinate are added to the second funnel.* Now both funnels 1 and 2 are

---

* In Figure 1 the upper phase (raffinate) stays in the same funnel and is, in effect, transferred to the one below it, while the lower phase (extractant) is moved on to the next funnel. The quantity of A transferred is indicated on the arrows.

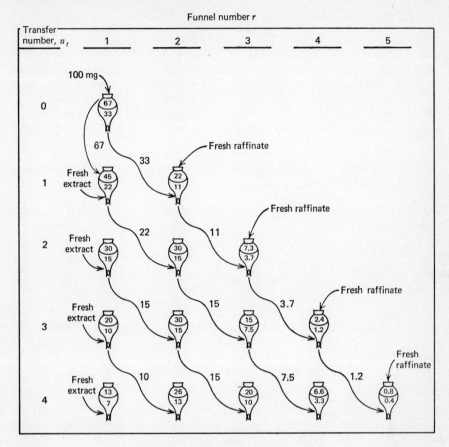

Figure 1. An example of stepwise countercurrent liquid–liquid extraction. Sample size 100 mg; $K_p = 2.0$; $\beta = 1.0$. The quantities shown inside the funnels are the equilibrium values based on the partition coefficient. The quantities shown on the arrows are the amounts transferred in each step.

shaken and allowed to come to equilibrium. The new amounts of A in each phase in each funnel are indicated in the figure. The rounding off of the numbers produce some ratios that are not exactly 2.0; for example, the amounts in funnel 1 in transfer 1 are 45:22. This process is repeated as shown in Figure 1, moving the extractant on to the next funnel and adding fresh extractant to funnel 1 and fresh raffinate to the newest funnel.

For a final example consider what occurs in funnel 2 in transfer 2. After transfer 1, 22.2 mg of A are in the raffinate (upper phase) in funnel 2 and will remain there for the next step, and 22.2 mg of A in the extractant are added from funnel 1. The total amount of A in funnel 2 in transfer 2 is 44.4 mg, and it is redistributed according to the partition coefficient:

$$(K_p)_A = 2.0 = \frac{29.6}{14.8} \tag{4}$$

The appropriate numbers are entered in Figure 1 after rounding off.

Table 1 summarizes the amount of A in the extractant phase in each funnel after each of four transfers (five equilibrations). The process can be carried on further, of course, but this is sufficiently far to observe what is happening.

Table 1. Amount of A in the Extractant Phase in Each Step of a Five-Funnel Countercurrent Extraction[a]

Number of Transfers, n_t	Funnel Number, r					Total Amount of A in Extractant (mg)
	1	2	3	4	5	
0	33.3					33.3
1	22.2	11.1				33.3
2	14.8	14.8	3.7			33.3
3	9.9	14.8	7.4	1.2		33.3
4	6.6	13.2	9.9	3.3	0.4	33.3

[a] $K_p = 2$, $\beta = 1$.

First, the amount of A in the raffinate in each funnel is twice the amount in the extractant (as is required by partition coefficient), and the total amount in each funnel is thus three times the amount in the extractant (the sum of the two phases). These values can be easily calculated from Table 1.

Second, in the last column in Table 1 the total amount of A in the extractant in each step is 33.3 mg. Obviously, then, the total amount of A in the raffinate is 66.6 mg, and the total amount is 100 mg (99.9 due to our rounding-off process). Although it may seem obvious to point out that the total amount of A remains at 100 mg, the equally obvious observation that the amount of A in the extractant remains constant at 33.3 mg is perhaps worth special note. This latter quantity corresponds in numerical value to Q, the fraction extracted:

$$Q = \frac{1}{K_p + 1} = \frac{1}{2 + 1} = \frac{1}{3} \tag{5}$$

and

$$\tfrac{1}{3} \times 100 \text{ mg} = 33.3 \text{ mg} \tag{6}$$

This observation suggests another useful definition of Q: it is the "fraction of solute that is in the extractant phase whenever the system is at equilibrium."

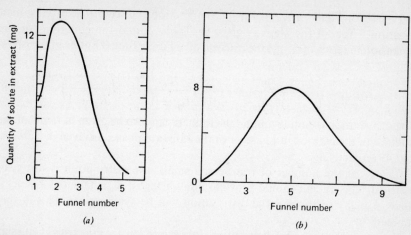

Figure 2. Distribution of solute in extraction funnels ($K_p = 2.0$): (a) after four transfers; (b) after 12 transfers.

Let us now look at the distribution of A in the five funnels after the fourth transfer. The data from Table 1 are plotted in Figure 2a. Note that we are using values for the amount of A in the *extractant*, although we could use the amount in the raffinate or the total amount equally well.

We can make three important observations:

1. The sample is now distributed in five funnels whereas it began in only one funnel; broadening has indeed occurred.

2. The distribution of the sample is nonsymmetrical.

3. The largest amount of sample is in the second funnel. Since all of the sample was in funnel 1 at the beginning, we can see that the sample is moving down the row of funnels, but slowly, since $K_p > 1$.

We examine these observations in more detail in the sections that follow.

Characterization of the Zones

Shape of Distribution. The nonsymmetrical distribution in Figure 2a is a plot of the expansion of the binomial expression

$$\left(\frac{K_p}{K_p + 1} + \frac{1}{K_p + 1}\right)^{n_t} \tag{7}$$

where n_t is the number of transfers. Equation 7 becomes quite obvious when it is noted that the first term, $K_p/K_p + 1$, is $(1 - Q)$, the fraction unextracted, and the second term, $1/K_p + 1$, is Q, the fraction extracted. The expansion of equation 7 for all the steps is given in Table 2, and the equation for the expansion in terms of $f_{(r,n_t)}$, the fraction of solute in funnel r after n_t transfers, is

$$f_{(r,n_t)} = \frac{n_t!}{(r-1)!\,(n_t - r + 1)!}\left(\frac{K_p}{K_p + 1}\right)^{n_t - r + 1}\left(\frac{1}{K_p + 1}\right)^{r-1} \tag{8}$$

The distribution of solute among the funnels can also be given in terms of the fraction in the extractant phase or in the raffinate phase, as given in Tables 3 and 4, respectively.

Clearly, if the number of transfers is small, the distribution is not symmetrical. The student should prove to himself that if K_p is taken as 1.0, as is done in many textbooks, the distribution will be symmetrical. This is very misleading.

However, the binomial distribution does approach a symmetrical Gaussian curve if the number of funnels is large. Returning to our example, this can be seen if more transfers and more funnels are considered. Such an expansion of Table 1 is given in Table 5, which shows the percentage of solute in each of 10 funnels for 56 transfer steps. Percentages less than 0.1 % are considered to be negligible.

The trend toward symmetry can be seen from Figure 2b, which shows the distribution after 12 transfer steps. The height of the band is less than that in Figure 2a since the sample is spread out into more funnels. That the distribution can be approximated by a Gaussian curve is well documented in theory and practice. Equation 8 can be put in the form of a Gaussian distribution (1):

$$f_{(n_t r)} = [2\pi n_t Q(1 - Q)]^{-\frac{1}{2}} \exp\left[-\frac{(r_{\max} - r)^2}{2n_t Q(1 - Q)}\right] \tag{9}$$

where $r_{\max}$ is the number of the funnel with the maximum amount of sample. It is the center of the solute band as it passes through the system. The quantity $(r_{\max} - r)$ is the number of funnels separating the one of interest, r, from the one with the maximum amount. The equation is valid only for a large number of transfer steps.

Extent of Zone Broadening. Equation 9 is seldom used to derive a given solute profile from a partition coefficient, or Q-value. The important point is that the distribution can be considered to be Gaussian and therefore that the width of the zone is proportional to the square root of n_t (2):

$$\sigma = [n_t Q(1 - Q)]^{\frac{1}{2}} \tag{10}$$

The important conclusion from the equation is that the band width, 4σ, is proportional to the square root of the number of transfer steps.

Table 2. Fraction of Solute in Each Funnel[a]

Number of transfers, n_t	Funnel Number, r				
	1	2	3	4	5
0	1	0	0	0	0
1	$\dfrac{K_D}{K_D+1}$	$\dfrac{1}{K_D+1}$	0	0	0
2	$\left(\dfrac{K_D}{K_D+1}\right)^2$	$2\left(\dfrac{1}{K_D+1}\right)\left(\dfrac{K_D}{K_D+1}\right)$	$\left(\dfrac{1}{K_D+1}\right)^2$	0	0
3	$\left(\dfrac{K_D}{K_D+1}\right)^3$	$3\left(\dfrac{1}{K_D+1}\right)\left(\dfrac{K_D}{K_D+1}\right)^2$	$3\left(\dfrac{1}{K_D+1}\right)^2\left(\dfrac{K_D}{K_D+1}\right)$	$\left(\dfrac{1}{K_D+1}\right)^3$	0
4	$\left(\dfrac{K_D}{K_D+1}\right)^4$	$4\left(\dfrac{1}{K_D+1}\right)\left(\dfrac{K_D}{K_D+1}\right)^3$	$6\left(\dfrac{1}{K_D+1}\right)^2\left(\dfrac{K_D}{K_D+1}\right)^2$	$4\left(\dfrac{1}{K_D+1}\right)^3\left(\dfrac{K_D}{K_D+1}\right)$	$\left(\dfrac{1}{K_D+1}\right)^4$

[a] $\beta = 1.$

Table 3. Fraction of Solute in Extractant Phase[a]

Number of Transfers, n_t	Funnel Number, r				
	1	2	3	4	5
0	$\dfrac{1}{K_D+1}$	0	0	0	0
1	$\left(\dfrac{1}{K_D+1}\right)\left(\dfrac{K_D}{K_D+1}\right)$	$\left(\dfrac{1}{K_D+1}\right)^2$	0	0	0
2	$\left(\dfrac{1}{K_D+1}\right)\left(\dfrac{K_D}{K_D+1}\right)^2$	$2\left(\dfrac{1}{K_D+1}\right)^2\left(\dfrac{K_D}{K_D+1}\right)$	$\left(\dfrac{1}{K_D+1}\right)^3$	0	0
3	$\left(\dfrac{1}{K_D+1}\right)\left(\dfrac{K_D}{K_D+1}\right)^3$	$3\left(\dfrac{1}{K_D+1}\right)^2\left(\dfrac{K_D}{K_D+1}\right)^2$	$3\left(\dfrac{1}{K_D+1}\right)^3\left(\dfrac{K_D}{K_D+1}\right)$	$\left(\dfrac{1}{K_D+1}\right)^4$	0
4	$\left(\dfrac{1}{K_D+1}\right)\left(\dfrac{K_D}{K_D+1}\right)^4$	$4\left(\dfrac{1}{K_D+1}\right)^2\left(\dfrac{K_D}{K_D+1}\right)^3$	$6\left(\dfrac{1}{K_D+1}\right)^3\left(\dfrac{K_D}{K_D+1}\right)^2$	$4\left(\dfrac{1}{K_D+1}\right)^4\left(\dfrac{K_D}{K_D+1}\right)$	$\left(\dfrac{1}{K_D+1}\right)^5$

[a] $\beta = 1$.

Table 4. Fraction of Solute in Raffinate Phase[a]

Number of Transfers, n_t	Funnel Number, r				
	1	2	3	4	5
0	$\dfrac{K_D}{K_D+1}$	0	0	0	0
1	$\left(\dfrac{K_D}{K_D+1}\right)^2$	$\left(\dfrac{1}{K_D+1}\right)\left(\dfrac{K_D}{K_D+1}\right)$	0	0	0
2	$\left(\dfrac{K_D}{K_D+1}\right)^3$	$2\left(\dfrac{1}{K_D+1}\right)\left(\dfrac{K_D}{K_D+1}\right)^2$	$\left(\dfrac{1}{K_D+1}\right)^2\left(\dfrac{K_D}{K_D+1}\right)$	0	0
3	$\left(\dfrac{K_D}{K_D+1}\right)^4$	$3\left(\dfrac{1}{K_D+1}\right)\left(\dfrac{K_D}{K_D+1}\right)^3$	$3\left(\dfrac{1}{K_D+1}\right)^2\left(\dfrac{K_D}{K_D+1}\right)^2$	$\left(\dfrac{1}{K_D+1}\right)^3\left(\dfrac{K_D}{K_D+1}\right)$	0
4	$\left(\dfrac{K_D}{K_D+1}\right)^5$	$4\left(\dfrac{1}{K_D+1}\right)\left(\dfrac{K_D}{K_D+1}\right)^4$	$6\left(\dfrac{1}{K_D+1}\right)^2\left(\dfrac{K_D}{K_D+1}\right)^3$	$4\left(\dfrac{1}{K_D+1}\right)^3\left(\dfrac{K_D}{K_D+1}\right)^2$	$\left(\dfrac{1}{K_D+1}\right)^4\left(\dfrac{K_D}{K_D+1}\right)$

[a] $\beta = 1$.

Table 5. Percentage of Solute in Extractant Phase[a]

Number of transfers, n_t	Funnel Number, r									
	1	2	3	4	5	6	7	8	9	10
0	33.3									
1	22.2	11.1								
2	14.8	14.8	3.7							
3	9.9	14.8	7.4	1.2						
4	6.6	13.2	9.9	3.3	0.4					
5	4.4	11.0	11.0	5.5	1.4	0.1				
6	2.9	8.8	11.0	7.3	2.7	0.6	0.1			
7	2.0	6.8	10.2	8.5	4.3	1.3	0.2			
8	1.3	5.2	9.1	9.1	5.7	2.3	0.6	0.1		
9	0.9	3.9	7.8	9.1	6.8	3.4	1.1	0.2		
10	0.6	2.9	6.5	8.7	7.6	4.6	1.9	0.5	0.1	
11	0.4	2.1	5.3	8.0	8.0	5.6	2.8	1.0	0.3	
12	0.3	1.5	4.2	7.1	8.0	6.4	3.7	1.6	0.5	0.1
13	0.2	1.1	3.3	6.1	7.7	6.9	4.6	2.3	0.9	0.2
14	0.1	0.8	2.6	5.2	7.1	7.1	5.4	3.1	1.3	0.5
15	0.1	0.6	2.0	4.3	6.5	7.1	6.0	3.8	1.9	0.7
16	0.1	0.4	1.5	3.6	5.8	6.9	6.4	4.5	2.6	1.1
17		0.3	1.2	2.9	5.0	6.5	6.5	5.1	3.2	1.6
18		0.2	0.9	2.3	4.3	6.0	6.5	5.6	3.9	2.1
19		0.1	0.6	1.8	3.6	5.5	6.4	5.9	4.4	2.7
20		0.1	0.5	1.4	3.0	4.9	6.1	6.1	4.9	3.3
21		0.1	0.4	1.1	2.5	4.2	5.7	6.1	5.3	3.8
22		0.1	0.3	0.9	2.0	3.7	5.2	5.9	5.6	4.3
23			0.2	0.7	1.6	3.1	4.7	5.7	5.7	4.7
24			0.1	0.5	1.3	2.6	4.2	5.4	5.7	5.1
25			0.1	0.4	1.0	2.2	3.7	5.0	5.6	5.3
26			0.1	0.3	0.8	1.8	3.2	4.5	5.4	5.4
27			0.1	0.2	0.6	1.5	2.7	4.1	5.1	5.4
28				0.2	0.5	1.2	2.3	3.6	4.8	5.3
29				0.1	0.4	1.0	1.9	3.2	4.4	5.1
30				0.1	0.3	0.8	1.6	2.8	4.0	4.9
31				0.1	0.2	0.6	1.3	2.4	3.6	4.6
32				0.1	0.2	0.5	1.1	2.0	3.2	4.2
33					0.1	0.4	0.9	1.7	2.8	3.9
34					0.1	0.3	0.7	1.4	2.4	3.5
35					0.1	0.2	0.6	1.2	2.1	3.2
36					0.1	0.2	0.5	1.0	1.8	2.8
37						0.1	0.4	0.8	1.5	2.5
38						0.1	0.3	0.7	1.3	2.2
39						0.1	0.2	0.5	1.1	1.9

[a] $K_p = 2$, $\beta = 1$.

Table 5. Percentage of Solute in Extractant Phase[a] (contd.)

Number of transfers, n_t	\	\	\	\	Funnel Number, r	\	\	\	\	\
	1	2	3	4	5	6	7	8	9	10
40						0.1	0.2	0.4	0.9	1.6
41						0.1	0.1	0.4	0.8	1.4
42							0.1	0.3	0.6	1.2
43							0.1	0.2	0.5	1.0
44							0.1	0.2	0.4	0.8
45							0.1	0.1	0.3	0.7
46								0.1	0.3	0.6
47								0.1	0.2	0.5
48								0.1	0.2	0.4
49								0.1	0.1	0.3
50								0.1	0.1	0.2
51									0.1	0.2
52									0.1	0.2
53									0.1	0.1
54									0.1	
55										0.1
56										0.1

[a] $K_p = 2$, $\beta = 1$.

Equation 10 can be given in terms of K_p instead of Q by making the substitutions given in Chapter 2:

$$\sigma = \frac{(n_t K_p)^{\frac{1}{2}}}{1 + K_p} \tag{11}$$

This equation shows the same square-root relationship, but it is valid only when $\beta = 1.0$. Grushka (3) goes one step further and points out that when the number of steps exceeds the number of funnels, $n_t > r$:

$$\sigma = (n_{max} K_p)^{\frac{1}{2}} \tag{12}$$

where n_{max} is the number of the transfer step that contains the maximum amount of solute. We shall return to this subject later; for the present it is the square-root relationship that is to be noted.

That is, if the number of transfers is doubled, the width of the zone becomes 1.4 times as large, $2^{\frac{1}{2}}$. This point is better illustrated in Table 6 by data calculated by Craig (1). Note that in each case, as the number of transfers (and funnels) is doubled, the number occupied goes up by $2^{\frac{1}{2}}$.

Table 6. Effect of Number of Transfers on Solute Distribution[a]

Number of Transfers	Funnels Occupied	
	Number	Percent
25	15	60
50	21	42
100	30	30
200	42	21
400	60	15

[a] After Craig (*1*).

Clearly, the process of countercurrent LLE results in a dilution of the sample—it is spread out into an increasingly large number of funnels, as column 2 in Table 6 indicates. This spreading is working against a possible separation and may even seem to preclude any separation. However, column 3 in Table 6 indicates that a smaller and smaller percentage of funnels is occupied as the number of transfers increases. Hence a given zone appears to be more narrow on a percentage basis. This can be seen from Figures 3 and 4,

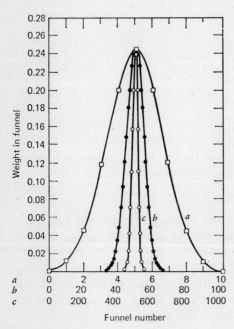

Figure 3. Comparative zone widths as a function of the number of funnels. Reprinted with permission from L. C. Craig, *Anal. Chem.* **22,** 1346 (1950). Copyright by the American Chemical Society.

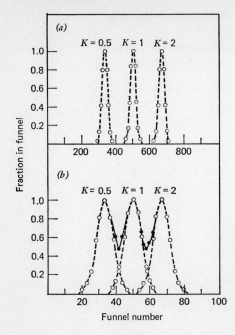

Figure 4. Calculated distribution patterns: (*a*) 1000 funnels; (*b*) 100 funnels; $K_p = 1.0$; $\beta = 1.0$. Reprinted with permission from L. C. Craig, *Anal. Chem.* **22,** 1346 (1950). Copyright by the American Chemical Society.

which are plotted on that basis. In practice, the apparent narrowing of zones is the more important feature of zone spreading; thus multistep processes are desirable in separations.

In summary, as the number of funnels and transfers increases, the number of funnels occupied increases according to the square root. The percentage of funnels occupied decreases.

Effect of Solvent Volume.   All the examples given so far have assumed equal volumes of extractant and raffinate. This need not be the case, although a large difference in the two volumes is experimentally impractical. The effect of the phase volume ratio β can be seen by considering the equation for Q derived in Chapter 2:

$$Q = \frac{1}{1 + K_p/\beta} \tag{13}$$

Recall that $\beta = V_{extr}/V_{raff}$; thus as β is increased Q, the fraction extracted must also increase. Table 7 lists some values for Q over a limited range of K_p and β values.

The effect of Q on zone broadening can be seen from equation 10 and Figure 5. The quarter-zone width σ has a maximum at $Q = 0.5$, which corresponds to the situation where $K_p = \beta$ (see Table 7). The most favorable circumstances (small σ) occur when Q is either small or close to 1.0.

Table 7. Effect of Phase Volume Ratio on Fraction Extracted

K_p	β	Q	k'
0.1	0.1	0.5	1.0
0.1	1.0	~1	0.1
0.1	10	~1	0.01
1.0	0.1	~0.1	10
1.0	1.0	0.5	1.0
1.0	10	~1.0	0.1
10	0.1	~0.01	100
10	1.0	~0.1	10
10	10	0.5	1.0

An alternative approach to this subject is via an examination of equations equivalent to equations 11 and 12 but expressed in terms of the partition ratio k' rather than the partition coefficient K_p. They are respectively

$$\sigma = \left[\frac{n_t k'}{(k' + 1)^2}\right]^{½} \tag{14}$$

and

$$\sigma = (n_{\max} k')^{½} \tag{15}$$

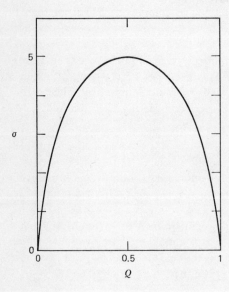

Figure 5. Effect of Q on zone broadening σ ($n_t = 100$).

In both cases the effect of β on σ is essentially the same as the effect of β on k'. Since k' is inversely proportional to β, as shown in Table 7, we can conclude that, as a first approximation, the zone width is inversely proportional to the square root of the phase volume ratio β. To keep σ small, one should use a large β, that is, a large ratio of extractant volume to raffinate volume. Table 7 shows that in such situations Q approaches unity which has been shown to be favorable. Still it must be repeated that β cannot vary too widely for practical reasons, and hence σ cannot be influenced very much by changing β.

Plate Theory of Chromatography

In their Nobel–prize paper of 1941, Martin and Synge (4) drew analogies among chromatography, distillation, and extraction. They considered the chromatographic column to consist of many LLE stages, or *theoretical plates*, a term taken from distillation terminology; this approach thus became known as the plate theory.

Although it may be instructive to visualize a chromatographic column as a series of stages or plates, like the separatory funnels just discussed, the plate theory is not very useful in chromatography. Nevertheless, it was a valuable contribution to the development of chromatographic theory in 1941, and the terms "theoretical plate" and "HETP" (height equivalent to a theoretical plate) remain in use. Many books contain detailed discussions of the development of the plate theory (5–8), so it is not included here. For a more useful and accurate description of chromatography we turn to the so-called rate theory. Before doing so, however, it will be useful to differentiate between a band and a peak.

The Distinction between Bands and Peaks

Up to this point it may have appeared that the terms "band," "peak," and "zone" have been used synonymously. However, we noted that the quarter-zone width was expressed in two slightly different ways (equation 11 versus equation 12). In the first case (equation 11) the sample was still completely contained *in* the system, the separatory funnels. When this is the case, the term "band" should be used to describe the solute concentration profile. In the second case (equation 12) transfers were continued after the solute arrived at the end of the system (the last funnel) and some sample was withdrawn with each transfer. We indicated this by stating that $n_t > r$ (whereas in the first case $n_t \lesssim r$). When the solute has been withdrawn from the system, the term "peak" should be used to describe the solute concentration profile. Thus the term "band" has a slightly different meaning from the term "peak," and sometimes the distinction between the two can be critical. The term "zone" will continue to be used as a more general term to include both bands and peaks.

Consider, for example, the data in Table 5. After 12 transfers, some solute is leaving the system with each additional transfer since there are only 10 separatory funnels. The *bands* after 4 and 12 transfers were shown in Figures 2a and 2b, respectively. Thus there are many bands that can be depicted as the solute passes through the system, and band broadening is directly proportional to the square root of the distance traveled (number of funnels), as just discussed.

By contrast, only one peak can be drawn for this solute. It is obtained by monitoring the sample as it elutes from the system. Thus the values in the last column of Table 5 represent the percentages of solute transferred out of the system on the next step. Some of the values are 0.1% in transfer 13, 0.2 in transfer 14, 0.5 in transfer 15, 0.7 in transfer 16, and so on. The peak maximum occurs in the last funnel on transfer 27 (actually transfer 26½), as can be seen from Table 5. The values from the last column are plotted in Figure 6a and represent the *peak* for this solute. Compare this peak with the *band* (shown in Figure 6b) representing the distribution of solute when the maximum is in the tenth funnel and assuming that the system is composed of more than 10 funnels. Obviously the peak and the band are quite different. As plotted in the figure, the leading edge of the band is on its right side, and the leading edge of the peak is just the opposite. When peaks and bands are asymmetrical, this distinction is important, as confusion of a band with a zone can lead to an erroneous conclusion about the type of asymmetry.

As a further distinction, consider the "apparent" effect of K_p on band broadening for bands and peaks. This time we shall assume that our system is

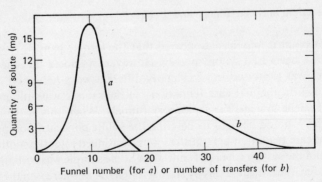

Figure 6. Differentiation between bands and peaks.

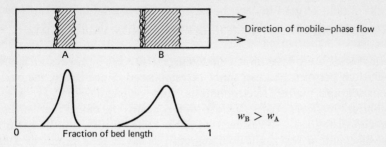

Figure 7. Chromatogram of bands for two solutes, A and B; $(K_p)_A > (K_p)_B$.

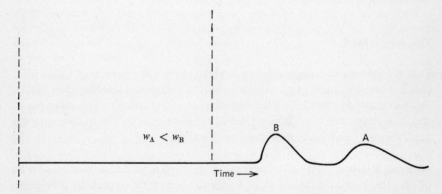

Figure 8. Chromatogram of peaks for the same solutes as in Figure 7.

a chromatographic column, shown in Figure 7. If we have two solutes A and B such that $(K_p)_A > (K_p)_B$, solute B will migrate faster than A (and be eluted first). At a given time, when both are still in the column, the relative *bands* are as shown in Figure 7. Solute B has moved *farther* than solute A so it is broadened more, $w_B > w_A$.

At a later time, when both solutes have been eluted from the column (B before A), the relative *peaks* are as shown in Figure 8. Now both have passed through the same length of column, but because solute A is retained more by the column, it has spent a longer *time* in the column, and consequently its peak is wider, $w_A > w_B$. Thus peak broadening depends on the *time* spent in the bed: the longer the time, the wider the peak. This distinction is amplified later in this chapter. In subsequent discussions care is taken to use the proper term in the proper context. When a distinction between terms is not necessary (and may in fact be detrimental), zone is used, although this is not as widely accepted as the other terms.

NONEQUILIBRIUM APPROACH (AS IN CHROMATOGRAPHY)

When one of the phases in a countercurrent process is always in motion, equilibrium between phases cannot be established. Since this is the case in chromatography, zone electrophoresis, and differential dialysis, a nonequilibrium model is needed. This section concerns the model that has come from chromatography. It has been called the "rate" theory, and credit for the original equation for gas chromatography is given to a group of Dutch scientists, including van Deemter, whose name is associated with the equation (9). A good historical summary of the concurrent development of the plate and rate theories has been given by Giddings (10).

Theoretical Basis

Two approaches to zone spreading in chromatography have been taken (11). One is a consideration of gross concentration changes as a solute zone moves through the system and is called *material* (or *mass*) *balance*. The other treats chromatography on a molecular level as a random stochastic process, and the random-walk model has been used to describe it.

Material Balance.   The distribution of solutes in a zone can be expressed as $\partial C/\partial t$, the partial differential of concentration with respect to time. The processes that cause spreading in a chromatographic system are related to diffusion, mass transfer, and the flow of the mobile phase. Hence the general mass balance equation can be written as (11)

$$\frac{\partial C}{\partial t} = D\,\frac{\partial^2 C}{\partial z^2} - u\,\frac{\partial C}{\partial z} + \sum_j (k_{ji}C_j - k_{ij}C) \tag{16}$$

where D is the diffusion coefficient, z is the direction of flow, and k is a proportionality constant. The solution of the equation yields the desired solute distribution function, but it is very complex (12, 13) and beyond the scope of this treatment. For the linear equations, which can be solved directly, the distribution is very close to a Gaussian curve. Thus the material balance model predicts the same distribution as the plate model.

Random-Walk Model.   As a single solute moves through a chromatographic bed, the individual molecules undergo a large number of sorption–desorption steps. When they are in the mobile phase, they move at the velocity of the mobile phase. When they are sorbed, they are immobile relative to the mobile phase. The probability that a given molecule will be mobile or sorbed is entirely random and permits us to use the random-walk model to approximate its behavior (*11, 14, 15*).

For simplicity we shall use a one-dimensional model, choosing our direction as the direction of the mobile-phase flow. If we assume that the distance between sorptions is constant, we can use l to represent this step length in our model. Steps in the forward and backward direction are equally probable. If a large number of molecules start out together on a random walk of N steps, they will gradually spread farther apart. The basic law of the random-walk model is

$$\sigma = lN_{rw}^{1/2} \tag{17}$$

where σ is the standard deviation or quarter-zone width of the resulting dispersion (as before), and N_{rw} is the number of steps. As applied to chromatography, we can conclude that zone spreading σ increases with the length of the steps, l, and with $N_{rw}^{1/2}$, not directly with N_{rw}. The latter effect, which may seem to be illogical, results from a cancellation effect of forward and backward steps.

In chromatography many random walks of different l and N_{rw} occur simultaneously. The combined effect is the sum of the variances, σ^2, of the individual steps:

$$\sigma_{\text{total}}^2 = \sigma_1^2 + \sigma_2^2 + \cdots + \sigma_i^2 \tag{18}$$

and the zone broadening resulting from several processes is equal to the *square root* of the sums of their variances:

$$\sigma_{\text{total}} = (\sigma_1^2 + \sigma_2^2 + \cdots + \sigma_i^2)^{1/2} \tag{19}$$

In chromatography, l represents the plate height H, and the product lN_{rw} is the distance migrated in the z-direction, L_z, as shown in Figure 9.

$$\sigma^2 = l^2 N_{rw} = HL_z \tag{20}$$

or

$$H = \frac{\sigma^2}{L_z} \tag{21}$$

which is the same result obtained by plate theory—zone broadening σ increases with the square root of the distance migrated.

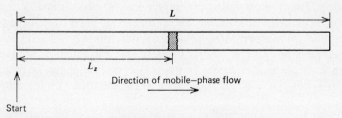

Figure 9. Illustration of the definition of L and L_z.

Equation 21 is commonly written as $H = \sigma^2/L$, where L is loosely defined as the length of the column or chromatographic bed. This can be misleading; the denominator in equation 21 should be the distance migrated, L_z, when referring to bands. For peaks it has a different interpretation; it is t_R^0 or V_R^0, the retention time or volume.* The choice between time or volume is made so that the denominator has the same units as the numerator. When using the equation $H = \sigma^2/L$, one must keep in mind the meaning of L for proper interpretation.

The important conclusions from the random-walk treatment are the following:

1. It has provided us with a new definition of H, the plate height: $H = \sigma^2/L_z$. Since there are no plates in a chromatographic column, it seems unwise to imply that there are by using the terms "plate" and "height equivalent to a theoretical plate" (HETP). However, there is at present no alternative to the use of these terms and symbols. Hence the best we can do is to refer to H as the "plate height" rather than as HETP. Further, we can visualize H as a measure of zone spreading σ^2 as a solute passes through a chromatographic bed.

2. It tells us that for bands the quarter-band width σ increases as the square root of the distance migrated, $L_z^{1/2}$.

3. For peaks σ increases as the square root of the retention time, $t_R^{1/2}$.

With this concept of H we can now proceed to devise a rate equation that relates H to column parameters. The basis for doing so is developed in Chapter 4.

Rate Equation

There is disagreement over the exact terms that should be included in a rate equation. This arises in part from attempts by different scientists to modify the basic theoretical approach with their empirical data. Basically, however, there are four contributions to zone spreading that should be included in a rate equation. They are presented in Chapter 4 as (1) eddy diffusion or convective mixing; (2) longitudinal molecular diffusion; (3) mass transfer in the mobile phase; and (4) mass transfer in the stationary phase.

Van Deemter and co-workers (9) derived their equation for *gas* chromatography; since mass transfer in gases (the mobile phase) is relatively fast, this term (H_{mtm}) was not needed in their equation. In its simplest form

---

* Note that the superscript zero has a specific meaning in gas chromatography, so it is used here in the interest of consistency in symbols.

the van Deemter equation is

$$H = A + \frac{B}{\bar{u}} + C\bar{u} \tag{22}$$

where $\bar{u}$ is the average linear velocity of the mobile phase, A is the term for eddy diffusion, $B/\bar{u}$ is longitudinal molecular diffusion, and $C\bar{u}$ is mass transfer in the stationary phase. The individual terms are given in expanded form in Table 8 along with several other rate equations. Each of these terms will be examined in more detail.

We can arrive at a satisfactory four-term rate equation covering the four most important causes of zone spreading by substituting the various expressions for σ derived in Chapter 4 into the definition of H derived from the random-walk model, $H = \sigma^2/L$. This equation is called the extended van Deemter equation in Table 8.

The first term is due to eddy diffusion or convective mixing:

$$H_{ed} = \frac{\sigma_{ed}^2}{L} = \frac{2\lambda\, d_p L}{L} = 2\lambda\, d_p \tag{23}$$

where d_p is the diameter of the particles in the bed and λ is a packing factor; the latter is determined by the range of particle sizes and how they are packed together in the bed. Typical values are given in Table 9. Clearly, band broadening is caused by large particles and poor (i.e., heterogeneous) packing.

The second term arises from longitudinal molecular diffusion:

$$H_{lmd} = \frac{\sigma_{lmd}^2}{L} = \frac{2\psi D_M t_M}{L} = \frac{2\psi D_M L}{\bar{u}L} = \frac{2\psi D_M}{\bar{u}} \tag{24}$$

where $\bar{u}$ is the average linear velocity of the mobile phase, D_M is the diffusion coefficient for the solute in the mobile phase, and ψ is an obstruction factor that expresses the obstruction the bed presents to free molecular diffusion. Typical values for ψ are given in Table 10 (16). Hawkes has recently shown that these values are flow dependent (17); his paper can be consulted for additional references and discussion. Note that since $\bar{u}$ is in the denominator in equation 24, the contribution of longitudinal molecular diffusion to band spreading will become very small at high velocities.

The third term in the rate equation arises from resistance to mass transfer in the stationary phase:

$$H_{mts} = \frac{\sigma_{mts}^2}{L} = \frac{q R_R(1 - R_R)\, d_f^2 \bar{u}}{D_S} \tag{25}$$

where d_f is the average thickness of the stationary phase, D_S is the diffusion coefficient for the solute in the stationary phase, and q is the configuration

Table 8. Comparison of Rate Equations

Reference	H_{ed}	H_{lmd}	H_{mts}	H_{mtm}	Other
van Deemter	$2\lambda d_p$	$\dfrac{2\gamma D_M}{\bar{u}}$	$\dfrac{8}{\pi^2}\dfrac{k' d_f^2 u}{(1+k')^2 D_S}$	—	—
Extended van Deemter	$2\lambda d_p$	$\dfrac{2\gamma D_M}{\bar{u}}$	$\dfrac{k'}{q(1+k')^2}\dfrac{d_f^2\bar{u}}{D_S}$	$\dfrac{\omega d_p^2\bar{u}}{D_M}$	—
Golay	—	$\dfrac{2D_M}{\bar{u}}$	$\dfrac{k'}{q(1+k')^2}\dfrac{d_f^2\bar{u}}{D_S}$	$\dfrac{1+6k'+11k'^2}{24(1+k')^2}\dfrac{d_p^2\bar{u}}{D_M}$	—
Giddings	—	$\dfrac{2\gamma D_M}{\bar{u}}+\dfrac{2\gamma_S D_S(1-R_R)}{\bar{u}R_R}$	$\dfrac{k'}{q(1+k')^2}\dfrac{d_p^2\bar{u}}{D_S}$	$\dfrac{\omega d_p^2\bar{u}}{D_M}$	$\left(\dfrac{1}{2\lambda d_p}+\dfrac{1}{H_{mtm}}\right)^{-1}$
Huber	$\dfrac{2\lambda d_p}{1+\lambda_2(D_M/\bar{u}\,d_p)^{1/2}}$	$\dfrac{a_d D_M}{\bar{u}}$	$a_S\dfrac{k'}{(1+k')^2}\dfrac{d_p^2\bar{u}}{D_S}$	$a_M\left(\dfrac{k'}{1+k'}\right)^2\dfrac{d_p^{3/2}(\bar{u})^{1/2}}{D_M^{2/3}}$	—

Table 9. Approximate Packing Factors

Mesh Range	λ
20–40	1
50–100	3
200–400	8

factor, which is determined by the configuration (or nature) of the stationary phase. A few values are given in Table 11.

The final term, for mass transfer in the mobile phase, is similar:

$$H_{\text{mtm}} = \frac{\sigma^2_{\text{mtm}}}{L} = \frac{\omega\, d_p^2 \bar{u}}{D_M} \tag{26}$$

where d_p is the average diameter of the particles in the bed, D_M is the diffusion coefficient for the solute in the mobile phase, and ω is a factor determined by the nature of the packing and approximately equal to 1.3. Both mass-transfer terms include $\bar{u}$ in the numerator, and thus they are dominant terms at high mobile-phase velocities. Further discussion about the effects of various parameters on H will be deferred until later. At this point we must take a closer look at the nature of flow patterns in chromatographic beds and try to explain the origin of the additional term in Giddings's equation (Table 8).

Flow through a real chromatographic bed is much more complex than our description in Chapter 4. In a column, for example, the velocity of the mobile phase is not uniform across the column. Thus an initial zone like that in Figure 10a becomes distorted as it moves down the column due to the slower movement of the mobile phase in the center of the column. The resulting distribution is shown in Figure 10b. This has two significant consequences: first, the σ of the zone is now much larger than expected, as shown in the

Table 10. Approximate Obstruction Factors[a]

Chromatographic Bed	ψ
Glass beads	0.6–0.7
Chromosorb W	0.74
Chromosorb P	0.46
Paper	0.7–0.9

[a] From Giddings (16).

**Table 11. Approximate Configuration
Factors**

Type of Bed	q
Uniform liquid film	$\frac{2}{3}$
Paper	$\frac{1}{2}$
Ion-exchange resin	$\frac{2}{15}$

figure; second, a concentration gradient now exists radially as well as longitudinally in the column. Thus diffusion will occur from x to y, as shown, and will tend to minimize the effect of the flow differences and thus help to decrease zone spreading.

In an analogous fashion, the same effect occurs in any channel in a packed bed—for example, in the space between two particles of solid support. Since these channels are numerous and randomly distributed, the actual diffusion in a packed bed is much more complicated than previously described. Giddings (18) tried to classify these effects according to the distances over which they are operative. He has suggested the following five types:

1. Transchannel—the unequal flow rates in a channel between two particles
2. Transparticle—the result of pools of stagnant *mobile* phase held in the particle

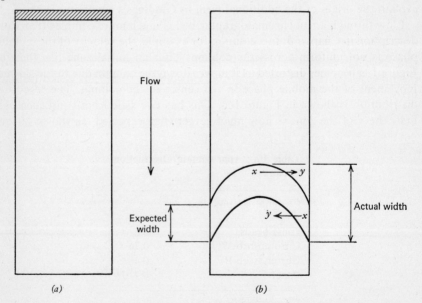

(a)   *(b)*

Figure 10. Typical band profiles in a chromatography column: *(a)* start; *(b)* later.

3. Short-range interchannel—the result of packing irregularities over the distance of a few particles

4. Long-range interchannel—the same as type 3, but operative over longer distances

5. Transcolumn—the overall effect in a column with which we began this discussion; also called the "wall effect"

As a further deviation from the classical model, Giddings proposed that the two effects, mass transfer in the mobile phase and eddy diffusion, are coupled and thus give rise to a new coupled term in the rate equation:

$$H_{\text{coupled}} = \Sigma \left(\frac{1}{H_{\text{ed}}} + \frac{1}{H_{\text{mtm}}} \right)^{-1} \tag{27}$$

He has also provided estimates of the mobile-phase packing parameters λ and ω for each type of flow inequality (18). These are given in Table 12 and supplement the values given earlier. Giddings's concept has been challenged, and the matter remains unresolved, but the simpler extended van Deemter equation will be sufficient for our purposes.

The equation by Golay in Table 8 was derived for GC columns that are "open" and not packed with a solid (19). Since these columns are not packed, they have very small diameters and have been called "capillary" columns, but the preferred name is "open tubular columns." The rate equation for open tubular columns has no eddy-diffusion term since there are no particles in the column to cause this type of zone broadening.

The equation by Huber in Table 8 was derived for liquid chromatography (20). It differs slightly from the van Deemter equation and is presented as an example of another worker's contribution. It will not be considered further here.

Table 12.   Approximate Values of the Mobile-Phase Packing Parameters[a]

Type of Velocity Inequality	ω	λ
Transchannel	0.01	0.5
Transparticle	0.1	10^4
Short-range interchannel	0.5	0.5
Long-range interchannel	2	0.1
Transcolumn	0.02–10	0.4–200
Transcolumn (coiled)	0–10^2	0–10^3

[a] From Giddings (18).

Table 13.   Comparison of k' and R_R

k'	R_R	$R_R(1 - R_R)$
1	0.5	0.25
4	0.2	0.16
9	0.1	0.09
49	0.02	0.0196

A comparison of equation 25 with the H_{mts} term in the extended van Deemter equation in Table 8 reveals one discrepancy. The equation contains the product $R_R(1 - R_R)$, and the table contains the quotient $k'/(1 + k')^2$. They are in fact equal and can be used interchangeably:

$$R_R(1 - R_R) = \frac{k'}{(1 + k')^2} \tag{28}$$

A comparison of k', and R_R is given in Table 13, from which one can see that as k' increases, $k'/(1 + k')^2$ decreases. This relationship is also shown in Figure 11.

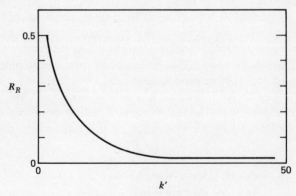

Figure 11.   Comparison of R_R and k'.

REFERENCES

1. L. C. Craig, *Anal. Chem.* **22,** 1346 (1950).
2. L. C. Craig and D. Craig, in *Technique in Organic Chemistry*, 2nd ed., Vol. 3, A. Weissberger (ed.), Wiley, New York, 1956, Chap. 11.
3. E. Grushka, *Separ. Sci.* **6,** 331 (1971).
4. A. J. P. Martin and R. L. M. Synge, *Biochem. J.* **35,** 1358 (1941).

5. A. I. M. Keulemans, *Gas Chromatography*, 2nd ed., Reinhold, New York, 1959, Chapter 4.

6. A. B. Littlewood, *Gas Chromatography*, Academic Press, New York, 1962, pp. 119–131.

7. H. A. Laitinen, *Chemical Analysis*, McGraw-Hill, New York, 1960, pp. 493–498.

8. I. M. Kolthoff, E. B. Sandell, E. J. Meehan, and S. Bruckenstein, *Quantitative Chemical Analysis*, 4th ed., Macmillan, New York, 1969, pp. 286–291.

9. J. J. van Deemter, F. J. Zuiderweg, and A. Klinkenberg, *Chem. Eng. Sci.* **5**, 271 (1956).

10. J. C. Giddings, *Dynamics of Chromatography*, Part I, Dekker, New York, 1965, pp. 13–25.

11. J. C. Giddings, *J. Chem. Educ.* **44**, 704 (1967).

12. J. F. K. Huber, in *Comprehensive Analytical Chemistry*, Vol. IIB, C. L. Wilson and D. W. Wilson (eds.), American Elsevier, New York, 1968.

13. J. C. Giddings, *Dynamics of Chromatography*, Part I, Dekker, New York, 1965, Chapter 4.

14. J. C. Giddings, *J. Chem. Educ.* **35**, 588 (1958).

15. J. C. Giddings, in *Chromatography*, 2nd ed., E. Heftmann (ed.), Reinhold, New York, 1967, Chapter 3.

16. J. C. Giddings, *Dynamics of Chromatography*, Part I, Dekker, New York, 1965, pp. 243–248.

17. S. J. Hawkes, *Anal. Chem.* **44**, 1296 (1972).

18. J. C. Giddings, *Dynamics of Chromatography*, Part I, Dekker, New York, 1965, pp. 40–65.

19. M. J. E. Golay, in *Gas Chromatography 1958*, D. H. Desty (ed.), Butterworths, London, 1958, p. 36.

20. J. F. K. Huber and J. A. R. J. Hulsman, *Anal. Chim. Acta* **38**, 305 (1967).

ZONE DISENGAGEMENT AND THE ACHIEVEMENT OF SEPARATION

9

In Chapter 8 equations were presented for the extent of zone broadening for multistep countercurrent processes. Two approaches were taken: the equilibrium approach, which is valid for countercurrent extraction, and the nonequilibrium approach, which is used primarily in chromatography. It was necessary to distinguish between "bands" and "peaks" in each case. To summarize the conclusions of Chapter 8, four equations are required for the four cases; these are given in Table 1. The table contains general equations that are independent of β, but the equations in Chapter 8 were valid only when $\beta = 1.0$. In each of the four cases it can be seen that the quarter-zone width σ is proportional to the square root of a parameter related to the system length. Specifically, these parameters are n_t, the number of transfers; L_z, the distance migrated by a solute band; n_{max}, the transfer number required to get

Table 1. Comparison of Equations for Expressing Zone Broadening and Migration Rates

A. Zone Broadening			
Bands		Peaks	
Equilibrium	Nonequilibrium	Equilibrium	Nonequilibrium
$\sigma = \left[\dfrac{n_t k'}{(1 + k')^2} \right]^{\frac{1}{2}}$	$\sigma = (HL_z)^{\frac{1}{2}}$	$\sigma = (n_{max}k')^{\frac{1}{2}}$	$\sigma = (t_R H)^{\frac{1}{2}}$

B. Zone Migration			
$r_{max} = n_t Q$	$\bar{v} = \bar{u}R_R$	$n_{max} = \dfrac{r - 1}{Q}$	$t_R = \dfrac{t_M}{R_R}$
	$S_{solute} = S_{solvent \atop front} R_F$	$= (r - 1)(1 + k')$	$= t_M(1 + k')$

the peak maximum into the last funnel; and t_R, the retention time for a chromatographic peak.

Before discussing the subject of separation of two solutes, we need to derive equations to express zone-migration rates for individual solutes—in other words, how fast a given solute travels through the system. We can then compare the distance between two solutes and their respective widths. For a complete separation the distance between the solute maxima has to exceed the 4σ width, as was shown in the definition of resolution in Chapter 2.

ZONE MIGRATION RATES

Equilibrium Approach

The rate at which a solute moves through a system obviously depends on its relative interactions with the two phases, which can be expressed as its partition coefficient K_p or partition ratio k'. To find the exact relationship, it will be helpful to examine the example given in Chapter 8 for a solute with $K_p = 2.0$. The data were given in Table 5, which gives the percentage of solute in the extractant phase in each funnel at each transfer. Since all of the solute does not travel together (it spreads out into an increasing number of funnels), it is customary to let the zone maximum represent the zone position.

At the start (transfer O) all of the solute is in the first funnel; by transfer 6 the zone maximum is in the third funnel; by transfer 9 it has moved to the fourth funnel, and so on. The zone position for the first 27 transfers is plotted in Figure 1, and the best straight line is drawn through the points; r_{max} is the number of the funnel containing the maximum amount of solute. The relationship is clear—the rate of zone migration is directly proportional to the number of transfers.

The exact relationship was first derived by Nichols (1) and was further elaborated by Grushka (2).* For bands it is

$$r_{max} = n_t Q \tag{1}$$

where r_{max} is the number of the funnel containing the band maximum, n_t is the number of transfers, and Q is the fraction extracted. Since Q is a function of the partition coefficient, equation 1 provides the fundamental relationship anticipated at the beginning of this section. The quantity Q can be viewed as the fraction expressing the relative rate of movement of the band maximum,

* Grushka's equations are given for $\beta = 1$ and for a partition coefficient K_p defined as the reciprocal of the K_p used in this book. His equations are changed to agree with the rest of this book.

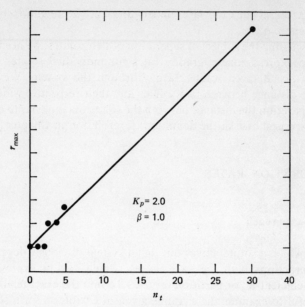

Figure 1.   Rate of movement of the band maximum for a solute with $K_p = 2.0$.

that is, $r_{\max}/n_t$. In our example, $K_p = 2$ and $Q = 1/(2 + 1) = 1/3$. The band maximum moves at a rate that is one-third of the transfer number, which is slower than the rate for a noninteracting solute ($K_p = 0$). This is because a K_p value greater than unity indicates that this solute is more strongly held by the raffinate (stationary) phase.

Application of equation 1 to the data in Table 5 of Chapter 8 reveals a small error. This results because n_t is so small and equation 1 is valid only for large n_t values.

For peaks Grushka has presented a slightly different equation (2):

$$n_{\max} = \frac{r - 1}{Q} \tag{2}$$

where $n_{\max}$ is the transfer number when the band maximum is in the last funnel (i.e., it is just becoming a *peak*). This equation holds for our example: $(r - 1) = (10 - 1) = 9$; $Q = 1/3$; and

$$n_{\max} = \frac{9}{1/3} = 27 \tag{3}$$

Table 5 in Chapter 8 shows that the zone maximum is in the last funnel in transfer 27. Equations 1 and 2 are included in Table 1.

Nonequilibrium Approach

The desired equation for the rate of zone migration in chromatography is inherent in the definition of the retention ratio R_F or R_R:

$$R_F = \frac{s_{\text{solute}}}{s_{\text{solvent front}}} \tag{4}$$

where s is the distance migrated and

$$R_R = \frac{\bar{v}}{\bar{u}} \tag{5}$$

$\bar{v}$ being the average velocity of the solute and $\bar{u}$ the average velocity of the mobile phase. When these two equations are rearranged, they are very similar to equation 1 and illustrate again that Q, R_R, and R_F are similar concepts:

$$s_{\text{solute}} = s_{\text{solvent front}} R_F \tag{6}$$

$$\bar{v} = \bar{u} R_R \tag{7}$$

These equations apply to bands.

The equation for peaks can also be derived from the definition of retention ratio. Equation 8 was derived in Chapter 2:

$$R_R = \frac{V_M^0}{V_R^0} \tag{8}$$

where V_M^0 is the volume of the mobile phase or the volume that would be required to elute a nonsorbed sample ($K_p = 0$) and V_R^0 is the retention volume for the solute. If the flow is constant, these volumes can be expressed as times:

$$R_R = \frac{t_M}{t_R} \tag{9}$$

where t_M is the time to elute a nonsorbed sample and t_R is the time to elute the solute of interest. Rearranging and substituting $R_R = 1/(1 + k')$, we get

$$t_R = \frac{t_M}{R_R} = t_M(1 + k') \tag{10}$$

The similarity between equation 10 and the one arrived at from the equilibrium approach can be seen in Table 1.

Summary

The zone migration rate has been expressed in four ways, which are summarized in Table 1. In each case the zone maximum moves in direct proportion to the "length" of the system, which is expressed by four different parameters: n_t, the number of transfers; $\bar{u}$, the average linear velocity of the mobile phase; $(r - 1)$, the number of funnels less one; and t_M, the retention time for a nonsorbed solute. The proportionality factor in each case is either Q, R_R, or R_F.

THE ACHIEVEMENT OF SEPARATION

A discussion about achieving a separation must involve at least two solutes, whereas our discussion to this point has been for individual solutes. If each of two solutes moves at a rate as indicated in Table 1, the distance between the maxima of the two solutes must also be proportional to the "length" of the system. To be specific, consider the equations derived for bands using the non-equilibrium approach (as in chromatography). The distance d between the band maxima for two solutes depends on their relative R_F values and the common distance the solvent front has moved, $s_{\text{solvent front}}$. Since the

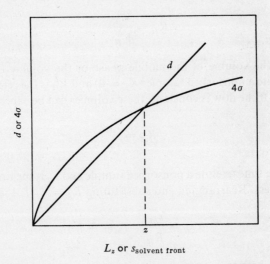

Figure 2. The achievement of separation. Reprinted from reference *3* by courtesy of Marcel Dekker, Inc.

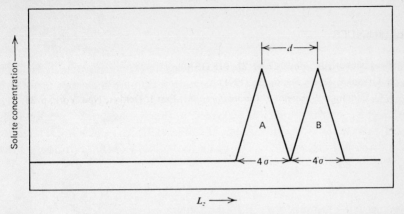

Figure 3. Concentration profiles for two bands at point Z in Figure 2 ($d = 4\sigma$).

solvent-front distance represents the length of the bed, d is directly proportional to it. On the other hand, the band width 4σ is directly proportional to the square root of the distance migrated, L_z.

These two equations are plotted in Figure 2 for a typical system. The parameters d and 4σ must be in the same units. Because one function is linear and the other depends on the square root, they must eventually cross, and that point is shown in the figure as z. At "lengths" greater than z, d is greater than the width 4σ and the two solutes are considered to be separated. This can be seen in Figure 3, in which the two bands are represented as triangles each with a base width of 4σ. Any further migration will cause d to become greater than 4σ (although both will increase), and the separation will improve. Hence it can be concluded that for any two solutes with a difference in R_F (or R_R or Q or K_p), a separation can be achieved if the system (chromatographic bed, number of countercurrent stages, etc.) is long enough. There is, of course, a practical limitation in length and time required (and pressure in column chromatography) that makes this generalization impractical in many cases. Still, the conclusion is basically true.

Finally, attention should be called to the definition of resolution R_S presented in Chapter 2. It is also based on a comparison of the two parameters d and σ:

$$R_s = \frac{2d}{(4\sigma)_A + (4\sigma)_B} \tag{11}$$

where the subscripts A and B refer to the two solutes. Where the two bands (or peaks) have the same width, that is, $(4\sigma)_A = (4\sigma)_B$, equation 11 reduces to

$$R_s = \frac{d}{4\sigma} \tag{12}$$

REFERENCES

1. P. L. Nichols, Jr., *Anal. Chem.* **22,** 915 (1950).
2. E. Grushka, *Separ. Sci.* **6,** 331 (1971).
3. J. C. Giddings, *Dynamics of Chromatography*, Part I, Dekker, New York, p. 33.

LIQUID–LIQUID EXTRACTION 10

The most common form of extraction is performed with two immiscible liquid phases. The sample is dissolved in the raffinate and is contacted with the extractant. In Chapter 2 the partition coefficient K_p was arbitrarily defined as

$$K_p = \frac{[\text{A}]_{\text{raff}}}{[\text{A}]_{\text{extr}}} \tag{1}$$

Two solutes can be separated if their partition coefficients differ, preferably one being greater than unity (and remaining in the raffinate) and the other less than unity (and being extracted).

Two types of application can be distinguished. Most common is the use of LLE to separate metal ions (usually involving some type of complexation) or complex organic mixtures. The other major application is the cleanup of reaction products to remove unwanted impurities. The first type is of major interest in analytical chemistry.

To a first approximation, one can think of the partition coefficient as the ratio of the solubilities of a given solute in the two liquids. However, this can be very misleading for several reasons. The solubilities of solids are related to their lattice energies, and these are not involved in LLE. The effect of the two liquid phases on each other and the possible formation of complexes with the solute cause differences in the system that are not the same as in simple solubility.

CLASSIFICATION OF METHODS

The term *exhaustive extraction* is used to describe the process of removing all of one substance from a sample. This would be the situation in the cleanup of a reaction product, for example, or in the separation of a given class of compounds. Usually such separations are performed by the crosscurrent process. The other type of extraction can be called *selective extraction* since it is applied to more difficult separations such as a mixture of metal ions. It is usually performed by the countercurrent process.

The following classification is based on the distinction between the crosscurrent and countercurrent processes. It is further subdivided according to the method of contacting the two phases: batchwise or continuous.

Batch Crosscurrent Extraction

Table 1 lists the equations for crosscurrent extraction derived in Chapter 2. Clearly, the fractions extracted and unextracted do not depend on the absolute quantity of solute, A. It is assumed, of course, that the amount of A does not exceed its solubility in both phases and form a third phase.

A plot of Q (in which β is assumed to be unity) is shown in Figure 1. Virtually no solute is extracted when $K_p \gtrsim 10^2$, while nearly all of the solute is

Table 1. Some Equations for Liquid–Liquid Extraction

Crosscurrent

$$K_p = \frac{(W_A)_{\text{raff}}}{(W_A)_{\text{extr}}} \times \frac{V_{\text{extr}}}{V_{\text{raff}}}$$

$$Q = \frac{V_{\text{extr}}}{V_{\text{extr}} + K_p V_{\text{raff}}}$$

$$= \frac{1}{1 + K_p/\beta} = \frac{\beta}{\beta + K_p}$$

$$= \frac{1}{1 + k'}$$

$$(1 - Q) = \frac{K_p V_{\text{raff}}}{V_{\text{extr}} + K_p V_{\text{raff}}}$$

$$= \frac{k'}{1 + k'}$$

Countercurrent

Bands	Peaks
$\sigma = \left[\dfrac{n_t k'}{(1 + k')^2} \right]^{1/2}$	$\sigma = (n_{\max} k')^{1/2}$
$r_{\max} = n_t Q$	$n_{\max} = \dfrac{r - 1}{Q} = (r - 1)(1 + k')$

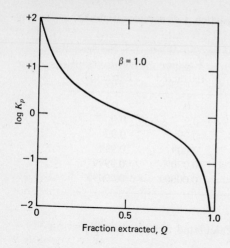

$\beta = 1.0$

Figure 1. Dependence of crosscurrent extraction efficiency on K_p

extracted when $K_p \lesssim 10^{-2}$. Such wide differences in K_p values are necessary if one-step extractions are to be useful.

When the partition coefficients are not that dissimilar, multiple extractions can be used. With each step the fraction Q is extracted and the quantity $(1 - Q)$ is unextracted. After n extractions, the fraction unextracted is

$$(1 - Q_n) = \left(\frac{K_p V_{\text{raff}}}{V_{\text{extr}} + K_p V_{\text{raff}}}\right)^n \tag{2}$$

and the fraction extracted is

$$Q_n = 1 - (1 - Q_n) = 1 - \left(\frac{K_p V_{\text{raff}}}{V_{\text{extr}} + K_p V_{\text{raff}}}\right)^n \tag{3}$$

Note that Q_n is *not* equal to $\left(\dfrac{V_{\text{extr}}}{V_{\text{extr}} + K_p V_{\text{raff}}}\right)^n$.

Example 1. In a five-step crosscurrent extraction, find the fraction extracted in each step and the total fraction amount extracted after each step. Assume $K_p = 0.1$, $V_{\text{raff}} = 10$ ml, and $V_{\text{extr}} = 9$ ml. In each step the fraction extracted, Q, is

$$Q = \frac{9}{9 + 10(0.1)} = \frac{9}{10}$$

In the first step 90% of the solute is extracted and 10% remains unextracted. In the second step 90% of the unextracted 10% is extracted, giving 9% (0.09) extracted for step 2. The fractions for each step are given in Table 2. The

Table 2.   Crosscurrent Extraction Calculations from Example 1

Step No.	Fraction Unextracted	Fraction Extracted	Total Fraction Extracted
0	1	0	0
1	0.1	0.9	0.9
2	0.01	0.09	0.99
3	0.001	0.009	0.999
4	0.0001	0.0009	0.9999
5	0.00001	0.00009	0.99999

fractions in the last column can be calculated directly from equation 3:

$$Q_1 = 1 - (1 - 0.9)^1 = 1 - 0.1 = 0.9$$
$$Q_2 = 1 - (0.1)^2 = 1 - 0.01 = 0.99$$
$$Q_3 = 1 - (0.1)^3 = 0.999$$
$$Q_4 = 1 - (0.1)^4 = 0.9999$$
$$Q_5 = 1 - (0.1)^5 = 0.99999$$

It is obvious that the numbers in the last column of Table 2 are approaching 1.0 asymptotically; consequently crosscurrent extractions are seldom carried beyond four or five steps. It is equally obvious that all of the solute (total fraction extracted of 1.0) cannot be extracted. However, for practical purposes a large fraction like 0.9999 can be considered to be a "complete" extraction. Note also that in crosscurrent-type extractions all of the extracts are combined, so that the total volume in example 1 is 45 ml after five steps.

Example 2. Compare the two-step extraction of example 1 with a one-step extraction using twice as much extractant (same total volume); that is, $V_{\text{extr}} = 18$ ml.

$$Q = \frac{18}{18 + 10(0.1)} = \frac{18}{19} = 0.95$$

In this example only 95% of the sample is extracted in one step, whereas 99% (Example 1) is extracted in two steps using the same total volume. From these examples we can generalize that multistep extractions are more effective than one-step extractions with the same total volume. This can be proved using equation 3.

Batch crosscurrent extractions are usually performed with simple separatory funnels.

Continuous Crosscurrent Extraction

Mathematical equations have not been worked out for continuous cross-current extraction. In the common laboratory procedures the volume of extractant is held constant and recycled. Just as in the stepwise process, the law of diminishing returns suggests that the extraction should not be continued for a long time. As the concentration of solute in the extractant increases, an equilibrium value is reached beyond which further extraction produces no change.

Two types of apparatus are required, depending on which phase is heavier. Figure 2 shows both types. The extractant is heated in the flask, vaporizes, and is condensed so that it contacts the raffinate phase as it returns to the reservoir.

Batch Countercurrent Extraction

Batch countercurrent extraction is used for selective extractions such as the separation of metal ions and complex organic mixtures. Many stages are usually used; the relevant equations were presented in Chapters 8 and 9. There are two modes of operation. In the first the number of equilibrations and transfers, n_t, is kept small enough so the sample components remain in the system, and their concentration profiles are called *bands*. In the second mode n_t is increased and the fractions coming off the end of the system are caught and separately kept. This method is referred to as *single withdrawal*, and the concentration profiles are called *peaks*. These equations are also included in Table 1.

The most popular apparatus is that designed originally by Craig (*1*), and the process is commonly referred to as Craig countercurrent extraction. The design of the individual tubes is shown in Figure 3. When both phases are present, the tube is rocked back and forth between positions *a* and *b*. When equilibration is complete, the phases are allowed to separate in position *b* so that the upper (extractant) phase flows through 2 into 3. The lower phase remains in 1. Finally, the tube is brought to position *c* and the extractant moves through 4 to the next tube.

Craig's original apparatus was made of glass, a typical one containing 200 tubes. Newer ones have a variety of shapes, are easier to clean, and can contain up to 1000 tubes. The entire process can be carried out automatically.

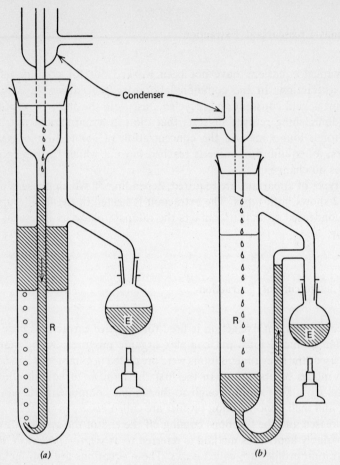

Figure 2. Apparatus for continuous liquid extraction (E = extractant; R = raffinate): (a) extractant lighter; (b) raffinate lighter.

MEASURES OF EFFICIENCY

Comparison of Crosscurrent and Countercurrent Processes

A simple example will serve to compare these two processes when used to attempt the separation of two solutes, A and B, by a two-step batch process. Assume that $(K_p)_A = 0.1$, $(K_p)_B = 10.0$, and $\beta = 1.0$. In the crosscurrent process the extractant from the first step contains 90.9% A and 9.1% B. In the second step fresh extractant is used on the raffinate (which contains 9.1% A

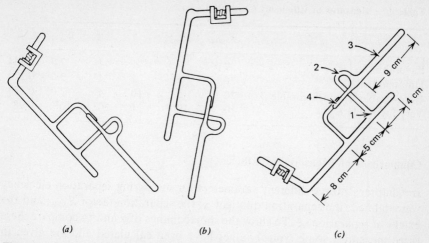

Figure 3. Craig countercurrent extraction funnels. For explanation of notation see the text. From *An Introduction to Separation Science*, by B. L. Karger, L. R. Snyder, and C. Horvath. Copyright by John Wiley & Sons.

and 90.9% B) to give $(90.9\% \times 9.1) + 90.9 = 99.2\%$ A and $(9.1\% \times 90.9) + 9.1 = 17.4\%$ B.

In the countercurrent process the first step yields the same result as crosscurrent. But in the second step each phase is reequilibrated with a *fresh* batch of the other phase. In the extractant, $90.9\% \times 0.909 = 82.6\%$ A and $9.1\% \times 0.091 = 0.83$ B.

The values of the fraction of the two solutes in the final extractant are given in Table 3. In crosscurrent extraction the recovery is high for both solutes, but there is a considerable amount of B present with the A. In countercurrent extraction the recovery is much lower, but the purity of A is much higher. Usually the latter is more important in analytical extractions.

Table 3. Comparison of Crosscurrent and Countercurrent Extraction Efficiencies

	Percentage in Extractant After Two Steps	
Sample	Crosscurrent	Countercurrent
A	99.2	82.6
B	17.4	0.83

Table 4. Measures of Efficiency Compared

	$(K_p)_B$	$(K_p)_A$	β	α	$S_{A/B}$	ξ
Favorable K_p	2.0	0.5	1	4	2.0	0.334
Unfavorable K_p	20	5	1	4	3.5	0.122
Unfavorable K_p but favorable β	20	5	10	4	2.0	0.334
Very favorable K_p	10^2	10^{-2}	1	10^4	10^2	0.98

Comparison of Measures of Efficiency

In Chapter 2, three different parameters for measuring separation efficiency were defined: the separation quotient α; the separation factor $S_{A/B}$; and the extent of separation, ξ. To show the shortcomings of α and to compare these three measures, some typical values have been calculated and are given in Table 4. Clearly, α is not an adequate measure of separation efficiency.

Definition of Theoretical Plate

Another measure of efficiency is the number of theoretical plates, n, and the related *height equivalent to a theoretical plate* (HETP), H. Grushka recently published equations for these two parameters for countercurrent distribution processes (2). His equations for bands and peaks are given in Table 5. It is interesting to compare these equations with those for chromatography.

Table 5. Definition of Number of Theoretical Plates, n, and HETP

	$\beta = 1.0$	$\beta \neq 1.0$
Bands	$n = \dfrac{r_{max}^2}{\sigma^2} = \dfrac{n_t Q}{1 - Q}$	$n = \dfrac{n_t}{k'}$
	$\quad = \dfrac{n_t}{K_p}$	
	$H = \dfrac{\sigma^2}{r_{max}} = \dfrac{K_p}{K_p + 1}$	$H = \dfrac{k'}{k' + 1}$
Peaks	$n = \dfrac{n_{max}^2}{\sigma^2} = (r - 1)\left(\dfrac{K_p + 1}{K_p}\right)$	$n = (r - 1)\left(\dfrac{k' + 1}{k'}\right)$
	$H = \dfrac{\sigma^2}{n_{max}} = K_p$	$H = k'$

THEORY

Solute–Solvent Interactions

A general introduction to intermolecular and interionic forces was given in Chapter 5. As noted there, the main forces in LLE are those in the bulk phase, which we have called *absorption forces*, and our discussion centers on solution theories.

A large part of the literature on LLE deals with metal extractions, where usually some type of complex is formed. An introduction to complex formation was given in Chapter 6. Morrison and Freiser (3) have published a monograph on LLE that serves as a good summary of the topic at that time. In it they classify metal complexes as being of three types: (1) coordination complexes, (2) chelate complexes, and (3) ion-association complexes. This follows the discussion in Chapter 6, with the addition of the third category, which covers uncharged compounds formed by the association of oppositely charged ions, either in pairs or larger groupings. Ion association is favored in solvents of low dielectric constant and thus is not prevalent in aqueous solutions. For further discussion and examples the monographs listed in the bibliography should be consulted.

Little attention has been given to the theoretical interpretation of extraction behavior. Noel and Meloan (4) have attempted empirical correlations between a number of systems and existing theories such as the Hildebrand solubility parameter. Their paper also provides a good bibliography of prior work. Two other recent papers attempt to show the relationship between the solubility parameter and ion-association systems (5, 6).

Phase-Volume Ratio

In our discussion of the separation quotient α as a measure of separation efficiency (Chapter 2) it was noted that it lacked usefulness because it failed to consider the *actual* values of the partition coefficients. In order to optimize the efficiency of an LLE, the phase volumes can be adjusted. This optimization is also related to the *actual* values of the partition coefficients as first suggested by Bush and Densen. Their equation was given in Chapter 2 as

$$\beta = \frac{V_{\text{extr}}}{V_{\text{raff}}} = [(K_p)_\text{A}(K_p)_\text{B}]^{1/2} \qquad (4)$$

This equation has recently been derived for LLE (7) and forms an important part of the theory of extraction.

PRACTICAL CONSIDERATIONS

Extraction Systems

Morrison and Freiser (3) present a good summary of the major extraction systems both by element and by complexing agent. For years a review on extraction was included in the biennial reviews appearing in *Analytical Chemistry*, but the last one was in 1968 (8) and the next one is expected in 1976: In the interim Katz (9) has reviewed the extraction publications in *Analytical Chemistry* covering 1968 to 1971.

One new procedure that has been suggested to speed up extractions is the formation of a homogeneous phase by raising the temperature of the system (10). On cooling, the two-phase system is reestablished and equilibrium is quickly attained.

Dimerization

Some solutes form dimers or higher oligomers in one or both of the solvents. When this occurs, the simple partition coefficient we have been using is not useful since all of the species need to be included in the equilibrium expression. For these situations the distribution coefficient K_D must be used (see Chapter 2):

$$K_D = \frac{\text{concentration of all forms of solute in raffinate}}{\text{concentration of all forms of solute in extractant}} \tag{5}$$

Similar situations exist when ion pairs and other complexes are formed, but only one example, acetic acid, will be discussed here.

When acetic acid (HOAc) is partitioned between water and benzene, two equilibria are established in addition to the simple distribution of the acid between the two phases. These are the formation of a dimer in the organic layer and the ionization in the aqueous layer. All of the reactions are summarized in the following diagram:

$$
\begin{array}{lll}
\text{Benzene layer} & 2\text{HOAc} \xrightleftharpoons{K_e} (\text{HOAc})_2 & \\
& \Big\updownarrow K_p & (6) \\
\text{Water layer} & \text{HOAc} + \text{H}_2\text{O} \underset{K_a}{\rightleftharpoons} \text{H}_3\text{O}^+ + \text{OAc}^- &
\end{array}
$$

The dimerization equilibrium constant K_c is

$$K_c = \frac{[(HOAc)_2]}{[HOAc]^2} \tag{7}$$

The ionization equilibrium constant K_a is

$$K_a = \frac{[H_3O^+][OAc^-]}{[HOAc]} \tag{8}$$

The distribution coefficient, which takes into account all forms in which the acetic acid is present (acid, dimer, acetate ion), is

$$K_D = \frac{[HOAc]_{\phi H} + 2[(HOAc)_2]_{\phi H}}{[HOAc]_{H_2O} + [OAc^-]_{H_2O}} \tag{9}$$

Substituting the expressions for K_c and K_a into K_D, the following relationship is obtained:

$$K_D = \frac{K_p(1 + 2K_c[HOAc]_{\phi H})}{1 + (K_a/[H_3O^+])} \tag{10}$$

From equation 10 we can see that the partition coefficient depends on the pH of the aqueous phase and on the amount of acetic acid in the benzene phase. Thus K_D depends on the total concentration of acetic acid.

Procedures

At least five considerations are involved in selecting the two liquid phases for an extraction. First, the two phases must be immiscible, although there is always some dissolution of each phase in the other; most commonly one phase is aqueous and the other nonaqueous. Second, the phases should be presaturated with each other before use to prevent changes in volume.

Third, the two phases must separate from each other quickly and without emulsion formation. Violent shaking of the two phases in the extraction vessel is not necessary and must be avoided or this action is likely to produce an emulsion. Sometimes the sample itself will serve as an emulsifying agent. To break an emulsion, try filtration, centrifugation, adding neutral salts to the aqueous layer, adding an additional solvent, changing the pH slightly, or altering the volume ratio (β) slightly. The volume ratio should be chosen to get the best separation as indicated by the Bush–Densen equation. Furthermore, crosscurrent extractions are more effective with several steps using a small volume of extractant rather than one large one.

A fourth consideration in the workup of the sample is whether or not the sample can be recovered easily from the extractant (or raffinate). Even if recovery is not required, detection of the solute and determination of its quantity may be necessary. For example, if the measuring step were to be performed by ultraviolet absorption, the extractant should be transparent in the ultraviolet.

Finally, we have to consider the rate at which the system comes to equilibrium. Fast reactions are desirable, but some substances, notably chelates like dithizonates, are slow in coming to equilibrium, and sometimes the rates are different in going from raffinate to extractant and vice versa.

Evaluation

Many separations of complex mixtures that could be performed by countercurrent LLE are now performed by chromatography, which is faster and simpler. However, there is no substitute for simple crosscurrent extractions, where the partition coefficients are highly favorable. Advantages and disadvantages are listed in Table 6.

Table 6. Evaluation of Liquid–Liquid Extraction

Advantages	Disadvantages
1. Simple	1. Multistep countercurrent processes are slow and laborious
2. Versatile	2. Theory inadequate for predicting new separations
3. Inexpensive	

REFERENCES

1. L. C. Craig, W. Hausmann, E. H. Ahrens, Jr., and E. J. Harfenist, *Anal Chem.* **23**, 1236 (1951).
2. E. Grushka, *Separ. Sci.* **6**, 331 (1971).
3. G. H. Morrison and H. Freiser, *Solvent Extraction in Analytical Chemistry*, Wiley, New York, 1957.
4. D. E. Noel and C. E. Meloan, *Separ. Sci.* **7**, 75 (1972).
5. H. Freiser, *Anal. Chem.* **41**, 1354 (1969).
6. T. Higuchi, A. Michaelis, and J. H. Rytting, *Anal. Chem.* **43**, 287 (1971).
7. E. Grushka, *Separ. Sci.* **7**, 293 (1972).
8. H. Freiser, *Anal. Chem.* **40**, 522R (1968).
9. S. Katz, *Am. Lab.*, p. 19, October 1972.
10. K. Murata, Y. Yokoyama, and S. Iheda, *Anal. Chem.* **44**, 805 (1972).
11. B. L. Karger, L. R. Snyder, and C. Horvath, *An Introduction to Separation Science*, Wiley, New York 1963, p. 112.

SELECTED BIBLIOGRAPHY

Fernando, Q., in *Separation Techniques in Chemistry and Biochemistry*, R. A. Keller (ed.), Dekker, New York, 1967, p. 357.

Freiser, H., *Critical Review in Analytical Chemistry* **1**, 47 (1970).

Irving, H., and R. J. P. Williams, in *Treatise on Analytical Chemistry*, Part I, Section C, I. M. Kolthoff and P. J. Elving (eds.), Interscience New York, 1961, Chapter 31.

Morrison, G. H. and H. Freiser, in *Comprehensive Analytical Chemistry*, Vol. 23, C. L. Wilson and D. W. Wilson (eds.), American Elsevier, New York, 1968.

Schweitzer, G. K. and W. van Willis, *Adv. Anal. Chem. and Instr.* **5**, 169 (1966).

von Metzsch, F. A., in *Physical Methods in Chemical Analysis*, Vol. IV, W. G. Berl (ed.), Academic Press, New York, 1961, pp. 317–456.

Weisiger, J. R. in *Organic Analysis*, Vol. IV, J. Mitchell, Jr. (ed.), Interscience, New York, 1954, p. 277.

CHROMATOGRAPHY: GENERAL

11

This chapter serves as a general introduction to the separation method called *chromatography*, which is important enough to require five chapters for adequate coverage. It includes the basic aspects of the theory, definitions, classifications, and common features and draws the basic distinction between the two major types, gas chromatography (GC) and liquid chromatography (LC). A historical introduction is not included, but it seems appropriate to call attention to the original 1906 paper by Tswett, which has been translated into English and is now readily available (*1*). It serves as a good introduction to the subject.

CLASSIFICATION OF METHODS

The main classification is according to the state of the mobile phase. Sub-classifications are based on the state of the stationary phase and the type of interaction between the solute and the stationary phase. These were presented in Chapter 2 and are summarized in Table 1. The first three LC methods under B in the table can be called "affinity" methods to distinguish them from B-4, exclusion chromatography, which is (ideally) based on size differences, and not affinity between the stationary phase and the sample. More often, however, the term "affinity chromatography" is applied to separations on activated gels in which reversible chemical bonds are formed, for example, in the separation of proteins (*2*).

The next classification is based on the method by which the sample migrates down the column: elution versus displacement. Elution is the more common method and the one we have concentrated on so far. In this process the sample is placed at one end of the stationary bed (the column) and washed with fresh solvent. Normally, in column chromatography, solvent (the mobile phase) is passed over the bed until all of the components in the sample have been "eluted." A detector at the column exit continuously records the components

Table 1. Classification of Chromatographic Methods

Stationary Phase	Name
A. Gas mobile phase; bed contained in a column:	
1. Solid, adsorbent	Gas–solid chromatography (GSC); adsorption chromatography
2. Solid, molecular sieve	Molecular-sieve chromatography
3. Liquid	Gas–liquid chromatography (GLC); absorption chromatography
B. Liquid mobile phase; bed contained in a column:	
1. Solid, adsorbent	Liquid–solid chromatography (LSC); adsorption chromatography; linear elution adsorption chromatography
2. Solid, ion-exchange resin	Ion-exchange chromatography
3. Liquid[a]	Liquid–liquid chromatography (LLC); absorption chromatography
4. Solid, molecular sieve	Molecular exclusion chromatography (gel-permeation chromatography)
C. Liquid mobile phase; bed is planar:	
1. Paper	Paper chromatography (PC)
2. Solid (adsorbent, ion-exchange resin, or molecular sieve)	Thin-layer chromatography (TLC)

[a] Normally the stationary phase is more polar than the mobile phase; if the opposite true, the process is called reverse-phase chromatography.

as they pass it, and the resulting chromatogram is usually displayed on a strip chart recorder (see Fig. 1).

In TLC and PC the sample solutes are usually not eluted, and the term "development" is used to describe the separation process. This is still elution chromatography, but the flow of mobile phase is stopped short of actual elution of the components from the bed. Finally, the term "gradient elution" is used to describe the elution process in which the mobile-phase composition is changed during a run. It is only applicable in LC, and the usual gradient is one of increasing "polarity," thus speeding up the elution of strongly retained components.

In the displacement procedure the sample is placed at one end of the bed as in elution. Usually the stationary phase is a solid that adsorbs the sample rather strongly, but the mobile phase is a substance that is sorbed by the stationary phase even more strongly than any of the sample components.

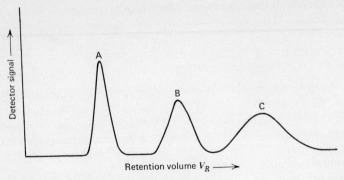

Figure 1. Typical elution chromatogram.

Obviously this is a type of LSC. The mobile phase displaces the sample from the bed and forces it ahead of it through the bed. If the components of the sample are sorbed to different extents, displacement will take place, so that each component displaces the one ahead of it that is less strongly sorbed. Consequently the sample component that is least strongly sorbed is the first one out of the bed. It is followed by those more strongly sorbed and finally by the mobile phase itself.

The type of chromatogram that results is shown in Figure 2. The three components of the sample are labeled: each zone contains only one component. The mobile-phase desorber is indicated by "D." Component A is shown coming off the column well ahead of component B to indicate that it is possible to have a component travel at a rate greater than that governed by

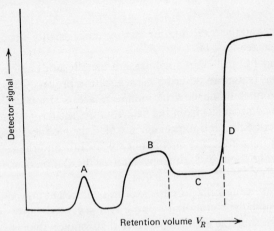

Figure 2. Typical displacement chromatogram.

displacement. Such a component would have very little affinity for the stationary phase. The length of each zone is proportional to its concentration. Since there is little overlap of components between zones and since the desorbent is usually a single liquid, the components of the sample can be recovered in pure form. If the detector measures a property of the components that is characteristic for each, then the height of each zone is characteristic of each for qualitative analysis. The refractive index is commonly used, and Figure 3 shows a chromatogram of a hydrocarbon mixture separated on a silica gel column using isopropyl alcohol as the desorber.

Though displacement methods are not too common, the fluorescent indicator adsorption (FIA) method for hydrocarbon types is a standard in the petroleum industry (3). A mixture of fluorescent dyes is added to the sample, one for each of the three types of hydrocarbons—paraffins, olefins, and aromatics. Then the mixture is placed on a silica gel column and developed using isopropyl alcohol as the displacer. Development is continued until the hydrocarbon types are separated, but before they are desorbed from the column. The presence of each type is identified by its fluorescence under an ultraviolet lamp, and the length of each band is measured. This value is correlated with the concentration of each type of component by comparing its length to the sum of all three bands. This is quite similar to the separation shown in Figure 3. Recently some attention has been given to improving this method in the light of modern LC techniques (5), and elution methods (6) will probably replace the displacement method for this analysis.

The last classification concerns the method of sample introduction as originally presented in Chapter 2. The two methods are zonal (batch) and frontal (continuous). The more common type is zonal, and it is the method that has been assumed in the discussions up to now. The sample is placed at the head of the bed all at once in a "spike" or narrow zone.

In the frontal analysis method the sample is added to the bed continuously at a constant rate along with the mobile phase. The sample components are sorbed by the bed until it becomes saturated, at which time the components "break through." Figure 4 shows a typical frontal analysis chromatogram for a three-component mixture. Component A breaks through first since the column becomes saturated with it first. When the column becomes saturated with component B, it breaks through and the effluent from the bed contains both A and B (as well as the mobile phase). Components B and C cannot be recovered in pure form by this technique, and the heights of the zones cannot be used for quantitative analysis. Frontal analysis has little value for analytical separations, but it does provide a good method for making thermodynamic measurements (7) and preparative-scale separations of the first component.

Unless otherwise noted, all subsequent discussion about chromatography will assume zonal elution operation.

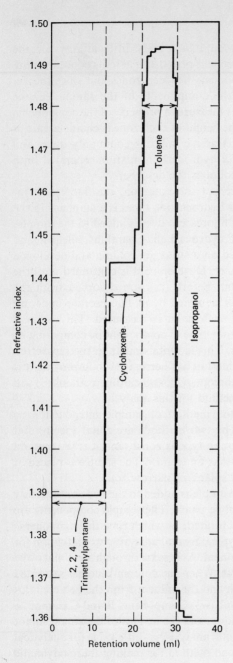

Figure 3. Actual displacement chromatogram. Reprinted with permission from B. J. Mair, *Ind. Eng. Chem.* **42,** 1355 (1954). Copyright by the American Chemical Society.

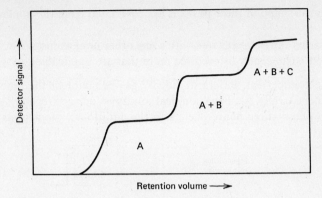

Figure 4. Typical frontal analysis chromatogram.

LINEARITY AND IDEALITY

In order to simplify theoretical discussions we usually assume that all systems are "ideal," and that is the assumption for most of this text. However, in chromatography some specific terms have come to be associated with "nonideal" behavior, and these need to be precisely defined.

From the beginning of GC the term "ideal" was used to denote the lack of zone broadening. It should be clear from the discussion in Chapter 8 that there is no such thing as ideal chromatography since some broadening is inherent in multistep processes. It is impossible to restrict the use of a general term like "ideal" to a specific meaning like the one just defined, so each time the term is used, the context will have to be considered.

The term "linear" is used for systems in which the partition coefficient remains constant as the concentration is changed. Linear and nonlinear isotherms were discussed in Chapter 5, where we saw that nonlinearity arises most often from adsorption rather than absorption.

The peak shapes that can result from each of the four possible combinations of these two parameters are shown in Figure 5 starting with the same sharp initial sample spike. In the linear-ideal case the shape of the eluted peak is the same as that for the sample "spike." In the linear-nonideal case the peak is broadened, but it is symmetrical. It represents the behavior we expect for good chromatographic systems and the type we shall ordinarily assume for our discussions. The nonlinear peaks depicted in Figure 5 show the effect of sample size. As can be seen, the peak maximum varies with sample size, which is highly undesirable. The mathematical equation that can be derived for the nonlinear-nonideal case is a second-order differential equation that cannot be solved, so the shape of these peaks is approximated from the nonlinear-ideal case.

Other terms can be associated with some other nonidealities. For example, Parcher (9) has recently listed three more that are applicable to GC:

1. Perfect–imperfect. Refers to "ideal" gas behavior of the mobile phase (see Chapter 3). In LC we have nonideal solutions such as "regular" solutions.

2. Two phase–three phase. Refers to the additional interactions that can

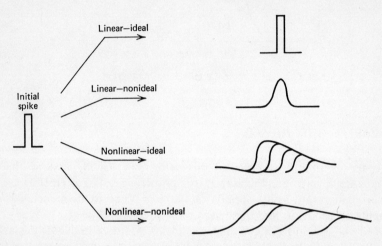

Figure 5. Representations of peak shapes for four types of chromatography.

occur at the gas–liquid interface and/or to adsorption by the solute on the solid support in GLC.

3. Dilute–nondilute. Refers to the fact that a sample has a finite volume that affects the flow rate when it is desorbed.

Although nonidealities such as these must be dealt with in practical chromatography, further elaboration of them is largely beyond the scope of this introductory presentation.

DEFINITIONS, EQUATIONS, AND THEORY

The fundamental theory of chromatography can best be given in a few simple equations as was done in Chapter 2. Two important ones are the definitions of the partition coefficient K_p and the retention ratio R_R. They are included in Table 2 along with all of the other important chromatographic equations. To summarize and review the definitions of the retention ratio R_R, it can be

Table 2. Summary of Important Chromatographic Equations

1. $(K_p)_A = \dfrac{[A]_S}{[A]_M}$

2. $K_p = k'\beta$

3. $\beta = \dfrac{V_M}{V_S}$

4. $k' = \dfrac{(W_A)_S}{(W_A)_M} = \dfrac{V_R'}{V_M} = \dfrac{V_N}{V_M^0} = \left(\dfrac{V_R^0}{V_M^0}\right) - 1 = \dfrac{1 - R_R}{R_R} = \left(\dfrac{1}{R_R}\right) - 1$

5. $R_R = \dfrac{V_M^0}{V_R^0} = \dfrac{\bar{v}}{\bar{u}}$

6. $V_R^0 = V_M^0 + K_p V_S$

7. $V_N = K_p V_S$

8. $R_R = \dfrac{V_M^0}{V_M^0 + K_p V_S}$

9. $V_R^0 = V_M^0(1 + k') = \dfrac{L}{\bar{u}}(1 + k') = n(1 + k')\dfrac{H}{\bar{u}}$

10. $(1 - R_R) = \dfrac{k'}{k' + 1}$

11. $R_R(1 - R_R) = \dfrac{k'}{(k' + 1)^2}$

12. $n = 16\left(\dfrac{t_R}{w}\right)^2 = \left(\dfrac{t_R}{\sigma}\right)^2$

considered to be: (a) the fraction of time a solute spends in the mobile phase, (b) the fraction of solute molecules in the mobile phase at any instant, and (c) the fraction of the mobile-phase velocity at which the solute moves.

The other fundamental chromatographic equation (equation 6 in Table 2) can be derived as follows.* When the maximum for a given solute has just reached the end of the column, one-half of it has been eluted in the volume we call the *retention volume* V_R^0, and the other half is still in the column. Thus the mass balance must be

$$V_R^0[A]_M = V_M^0[A]_M + V_S[A]_S \tag{1}$$

Dividing both sides of this equation by $[A]_M$, we get

$$V_R^0 = V_M^0 + V_S \dfrac{[A]_S}{[A]_M} \tag{2}$$

* For a more rigorous derivation see Karger, Snyder, and Horvath (10).

and since $[A]_S/[A]_M$ is the partition coefficient,

$$V_R^0 = V_M^0 + K_p V_S \qquad (3)$$

When the mobile-phase volume V_M^0 is subtracted from the total retention volume for a given solute, the difference is called the *net retention volume V_N*:

$$V_R^0 - V_M^0 = V_N = K_p V_S \qquad (4)$$

This important equation provides the link between the experimentally measured retention volume and the main thermodynamic parameter, K_p. As would be expected, as K_p increases, the retention volume increases.

Finally, the theory can be summarized by reference to the problem of separating two solutes. As they pass through the column, each interacts with the stationary phase to a degree determined by its partition coefficient (see Chapter 5) and elutes from the column at a time (or volume) reflecting this interaction. This retention time (or volume) can best be expressed as V_N (see equation 4). If the two solutes have different partition coefficients, they will elute at different times (in different volumes). The other aspect of the process that determines whether or not the solutes have been separated is the broadening of the zones as they pass through the system (Chapter 8). This broadening is measured as the number of theoretical plates, n, or the plate height H (which will be defined for chromatography in the next section). The achievement of separation is a comparison of the separation of zone centers and zone broadening (Chapter 9).

Bed Efficiency

The number of theoretical plates, n, is defined as

$$n = \left(\frac{t_R}{\sigma}\right)^2 = 16\left(\frac{t_R}{w}\right)^2 \qquad (5)$$

where t_R, w, and σ must be measured in the same units. Usually this calculation is performed on a peak like that shown in Figure 6, and t_R and w are distances along the x-axis and are measured in millimeters.

It must be remembered that since actual "plates" do not exist in a chromatographic bed, n simply represents a measure of the efficiency of the bed. It is a relative measure of the broadening (w) that has occurred while the sample passed through the system (t_R). It should be noted that the same terms, n and H, are used in distillation, but that comparisons of efficiencies of the two techniques cannot be made by comparing plates or plate heights. It takes many more chromatographic plates to achieve a given separation by GC

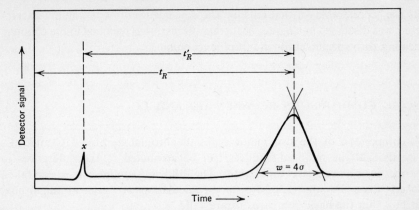

Figure 6. Figure used to define n, the number of theoretical plates ($x =$ nonretained component).

than it does to achieve the same separation by distillation (*11*). This is another limitation of the theoretical plate concept.

In Figure 6, a small peak x is included to represent a solute that is not sorbed and elutes in a time equivalent to the dead volume in the bed, V_M or t_M. Since all solutes must pass through this same volume, there are good theoretical reasons for eliminating it from consideration. (This was done in equation 4 to get the direct relationship between K_p and retention volume.) Hence the efficiency can also be defined in terms of the "effective" number of theoretical plates, n_{eff}:

$$n_{\text{eff}} = 16\left(\frac{t'_R}{w}\right)^2 = \left(\frac{k'}{k' + 1}\right)^2 n \qquad (6)$$

in which the retention time t'_R is measured from the maximum of the non-retained solute to the maximum of the main solute peak (Fig. 6). This retention time is designated by a prime and is called the *adjusted retention time*. As noted in equation 6, the term $(k'/k' + 1)^2$ relates n_{eff} to n. Although n_{eff} is not used as often as n, it is a better indication of the efficiency of a bed.

The other measure of bed efficiency is the plate height H, which is calculated from n:

$$H = \frac{L}{n} \qquad (7)$$

where L is the length of the bed. Unlike n, H is independent of the length, and its use is preferable for that reason. The dependence of H on column parameters was discussed in Chapter 8; further discussion is included in the chapters dealing with specific chromatographic techniques.

SOME COMPARISONS BETWEEN GC AND LC

A comparison of the two major types of chromatography, GC and LC, reveals that the major differences can be attributed to the differences in properties between gases and liquids. The following summary has been taken largely from Giddings (*12*). Most of the calculations are order-of-magnitude, at best, but the basic comparisons are valid.

First of all, GC is limited to the analysis of volatile substances or volatile derivatives of substances. This is probably the most important factor in choosing one technique versus the other for an analysis. The upper limit of volatility for GC is variable, but GC is seldom used for compounds (or derivatives) that boil above 450°C.

A second important difference is that gases normally used in GC are inert and have no affinity for the solute. Giddings (*12*) says that the "density of intermolecular attraction" is much lower—about 10^4 times—in gases than in liquids. The gas does not cause the solute to desorb—it simply carries it down the column; the mobile liquid in LC competes actively with the stationary phase for the solute molecules. It is for this reason that we have previously compared GC to extractive distillation (where vapor pressure is the important parameter) rather than to extraction (where solubility is the important parameter).

The other comparisons deal more with operating conditions than with major distinctions between the two types of chromatography. These are as follows:

1. The diffusivity is less in liquids by a factor of about 10^5. This has a large effect on the speed at which LC can be carried out.

2. The viscosity of liquids is greater by a factor of about 10^2.

3. The surface tension of liquids is greater by about 10^4. (Gases can be considered to have no surface tension.) The surface tension of liquids makes possible ascending TLC and PC due to the driving force of capillarity.

4. The density of liquids is greater by about 10^3.

5. Liquids are virtually incompressible, and gases are readily compressed. Consequently GC parameters that relate to the volume of mobile gas must be corrected to average values that compensate for the variation in volume with pressure.

Efficiency and Speed

To compare efficiencies, Giddings (12) has derived the following relationship for the limiting number of theoretical plates, $n_{\lim}$:

$$\frac{(n_{\lim})_{LC}}{(n_{\lim})_{GC}} = \frac{\eta_G}{\eta_L} \frac{D_G}{D_L} \tag{8}$$

We have already noted that the ratio of viscosities η_G/η_L is 10^{-2} and that the ratio of diffusion coefficients is 10^5. Therefore the limiting number of theoretical plates is 10^3 times greater in LC:

$$\frac{(n_{\lim})_{LC}}{(n_{\lim})_{GC}} = 10^{-2} \times 10^5 = 10^3 \tag{9}$$

One approximate calculation of the maximum number of plates in LC, $(n_{\lim})_{LC}$, is 10^8 plates, regardless of the bed length (12). This is much higher than any reported experimental data.

The derivation of equation 8 assumed that the pressure drop across a column (ΔP) is the same in GC and in LC.* However, D_G is pressure dependent and D_L is not, so this assumption is not valid. Consequently we should define a hypothetical gas diffusion coefficient that is independent of pressure, D'_G:

$$D_G = \frac{3}{2} \frac{D'_G}{\Delta P} \tag{10}$$

Making this substitution in equation 8 yields

$$\frac{(n_{\lim})_{LC}}{(n_{\lim})_{GC}} = \frac{3}{2} \frac{\eta_G D'_G}{\eta_L D_L \Delta P} \approx \frac{10^3}{\Delta P} \tag{11}$$

where ΔP is in atmospheres. The interesting conclusion that can be drawn from this equation is that an increase in pressure of about 10^3 atmospheres should make GC as efficient as LC. This high pressure is not the normal operating pressure for conventional GC, of course. In fact, the use of very high pressures usually results in the formation of supercritical fluids (see next section).

Although LC is predicted to be the more efficient technique, it is also the slower one. Consequently a realistic comparison of LC and GC should involve comparisons of both efficiency and speed. A theoretical comparison has been made and is shown in Figure 7 (13). Note that this is a log–log plot of

---

* The phrase "pressure drop across a column" refers to the pressure differential between the inlet and the outlet of the column, not across its cross section.

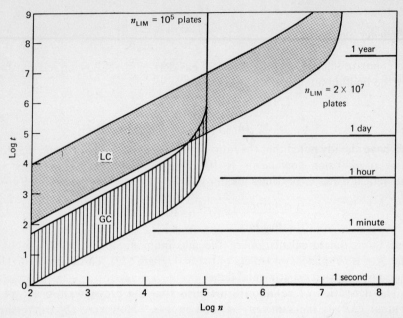

Figure 7. Comparison of GC and LC efficiencies and analysis times. Reprinted with permission from L. C. Giddings, *Anal. Chem.* **37**, 60 (1965). Copyright by the American Chemical Society.

time versus the number of plates. For small values of n, GC will operate faster (smaller t). However, for separations requiring more than 10 plates, only LC will do the job. It is only fair to note that the time required to achieve a large n in LC is very long!

In summary, we have found that GC is faster than LC, but that LC has the potential for performing more difficult separations. Consequently modern LC is often run at high pressures even at the expense of decreasing the efficiency.

Fluid Chromatography

If a chromatographic system is operated at very high pressures, the mobile phase can be above its critical pressure and have the properties of a gas and a liquid. It can be called a supercritical fluid, and the technique is called *fluid chromatography* (FC). The extension of GC into high-pressure FC is desirable according to equation 11 if one wants to increase the efficiency of the system. This suggestion was made in 1964 by Giddings, and in the next few years several groups began experimental work at pressures of 1000 atmospheres and higher (*14–17*). Although these high pressures are dangerous and the equipment is expensive and troublesome, FC does offer the advantages of speed and extremely high selectivity (as the pressure is varied) over LC. The big advantage over GC is that nonvolatile substances can be chromatographed at a relatively low (usually less than 200°C) temperature. Some of the fluids that have been used are ammonia, carbon dioxide, isopropanol, pentane, and some fluorocarbons. No commercial apparatus has appeared on the market, and it appears that the use of FC will remain quite limited.

Reduced Parameters

Some comparisons of GC and LC are facilitated if they are made on the basis of the following reduced parameters: the reduced plate height h

$$h = \frac{H}{d_p} \tag{12}$$

and the reduced velocity v

$$v = \frac{d_p}{D_M} v \tag{13}$$

where d_p is the particle diameter of the solid support and D_M is the diffusion coefficient for a solute in the mobile phase.

One such comparison has been reported by Knox and Saleem (*18*) in which they plotted $\log h$ versus $\log v$. Their plots do show the validity of using reduced parameters and suggest means for optimizing conditions in GC and LC. Even so, other workers are less enthusiastic about the use of reduced parameters, and they have not been widely adopted.

COMMON FEATURES

Instrumentation

The apparatus required for plane chromatography (PC and TLC) is minimal, and this discussion applies only to column chromatography. The major parts of a chromatographic instrument are listed in Table 3 and shown schematically in Figure 8. Detectors for both GC and LC have a number of parameters in common. Most of them have been recently summarized by Keller (*19*).

Other instrumental variables that are common to GC and LC are "dead volume" in the mobile-phase stream, which should be kept to a minimum

Table 3. Components of a Column Chromatograph

| Component | Specific Item in | |
	GC	LC
1. Pressurized mobile phase	Compressed gas	Liquid pump
2. Flow control	Valve	Valve
3. Temperature control	Oven, air bath	Oven or water-jacket
4. Sample introduction	Syringe or gas-sampling loop	Syringe, loop, or "stop-flow"
5. Column	Metal or glass tube	Metal or glass tube
6. Detector	TCD, FID, etc.	UV, RI, etc.
7. Recorder, strip chart	1 mV, 1 sec	10 mV, 1 sec
8. Other	Temperature programmer	Pulse damper, precolumn for LLC, gradient former

(*20*); the ratio between the particle size and the column diameter, which should be optimized (*21*); the effect on column efficiency of coiling the column (*22*); and the sample introduction system, the volume of which should be kept small.

Sampling

The sample should always be placed on the bed "as a spike" or, in other words, "in as narrow a zone as possible." For liquid samples introduced with a syringe this means a fast injection. In GC it means that the liquid must be rapidly volatilized (a heated injection zone is required) or the sample must be injected directly onto the column. In TLC and PC the sample is placed on the bed successively using small amounts and allowing the solvent to dry between applications. This keeps the sample spot small.

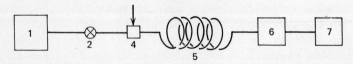

Figure 8. Schematic diagram of a chromatograph. Numbers refer to the components listed in Table 3.

Packing Columns*

Highest column efficiencies are obtained when columns are tightly packed with small particles of a narrow range of sizes. Due to the different characteristics of the solid supports and the fact that LC columns are usually packed with smaller particles than are used in GC, there is no unique method. Many satisfactory procedures exist, and some are given in references *23* through *27*.

SPECIAL TECHNIQUES

Support-Free Chromatography

One of the main differences between LLE and LLC is that in the latter technique the stationary liquid is held on a solid support. If this were not the case, it would be difficult to keep the mobile liquid from washing it out of the system. The solid support also provides a high surface area, which is desirable for fast, continuous analysis.

Recently, however, no less than four different methods have been proposed for performing LLC without a solid support (*28*). These new techniques have been called *countercurrent chromatography*, and they represent a technique that combines LLE and LLC. In general, they provide a high-efficiency separation without the undesirable adsorption problems that solid supports sometimes present.

Somewhat similar is a continuous GC method that does not use a packed bed (*29*). It consists of a pair of parallel disks on which the stationary liquid is coated. The sample enters at the center and both disks are rotated.

Centrichromatography

Centrichromatography is an LC technique in which the mobile liquid is moved through a conventional column by centrifugal force (*30–32*). A commercial apparatus is available, but the method has not become popular.

REFERENCES

1. H. H. Strain and J. Sherma, *J. Chem. Educ.* **44,** 238 (1967).
2. R. H. Reiner and A. Walch, *Chromatographia* **4,** 578 (1971); W. G. Gelb, *Am. Lab.* **5** [10], 61 (1973).
3. American Society for Testing and Methods, *1968 Book of ASTM Standards*, Part 17,

* Open tubular columns are not "packed" and are an exception to this discussion. They are discussed in Chapter 12.

p. 506, Method D-1319-66T; D. W. Criddle and R. L. Le Tourneau, *Anal. Chem.* **23**, 1620 (1951).

4. B. J. Mair, *Ind. Eng. Chem.* **42**, 1355 (1954).

5. T. A. Washall, *Anal. Chem.* **41**, 971 (1969).

6. R. E. Robinson, R. H. Coe, and M. J. O'Neal, *Anal Chem.* **43**, 591 (1971).

7. R. Stevenson, *J. Chromatogr. Sci.* **9**, 257 (1971).

8. C. J. Chen and J. F. Parcher, *Anal. Chem.* **43**, 1738 (1971).

9. J. F. Parcher, *J. Chem. Educ.* **49**, 472 (1972).

10. B. L. Karger, L. R. Snyder, and C. Horvath, *An Introduction to Separation Science*, Wiley, New York, 1973, p. 166.

11. J. J. van Deemter, F. J. Zuiderweg, and A. Klinkenberg, *Chem. Eng. Sci.* **5**, 271 (1956).

12. J. C. Giddings, *Dynamics of Chromatography*, Part I, Dekker, New York, 1965, pp. 293–301.

13. J. C. Giddings, *Anal. Chem.* **37**, 60 (1965).

14. J. C. Giddings, M. N. Meyers, and J. W. King, *J. Chromatogr. Sci.* **7**, 276 (1969).

15. M. N. Meyers and J. C. Giddings, in *Separation Techniques in Chemistry and Biochemistry*, R. A. Keller (ed.), Dekker, New York, 1967, p. 121.

16. N. M. Karayannis, A. H. Corwin, E. W. Baker, E. Klesper, and J. A. Walter, *Anal. Chem.* **40**, 1736 (1968).

17. R. E. Jentoft and T. H. Gouw, *Anal. Chem.* **44**, 681 (1972).

18. J. H. Knox and M. Saleem, *J. Chromatogr. Sci.* **7**, 745 (1969).

19. R. A. Keller, *J. Chromatogr. Sci.* **11**, 223 (1973).

20. See, for example, V. Maynard and E. Grushka, *Anal. Chem.* **44**, 1427 (1972).

21. J. C. Sternberg and R. E. Poulson, *Anal. Chem.* **36**, 1492 (1964).

22. H. Barth, E. Dallmeier, and B. L. Karger, *Anal. Chem.* **44**, 1726 (1972).

23. S. T. Sie and N. van den Hoed, *J. Chromatogr. Sci.* **7**, 257 (1969) (for LC).

24. *Whatman Data Sheet* 2, 1968 (for IE celluloses).

25. Applied Science Laboratories, *Gas-Chrom Newsletter* **11**[3], 1 (1970) (for GC).

26. C. Pidacks, *J. Chromatogr. Sci.* **8**, 618 (1970) (for gel chromatography).

27. L. R. Snyder, *J. Chromatogr. Sci.* **7**, 352 (1969) (for LC).

28. Y. Ito and R. L. Bowman, *Anal. Chem.* **43**[13], 69A (Nov. 1971); *J. Chromatogr. Sci.* **8**, 315 (1970).

29. M. V. Sussman and C. C. Huang, *Science* **156**, 974 (1967).

30. F. W. Karasek, *Research and Development*, p. 43 (July 1970).

31. E. Rib, S. C. Harris, J. Matsumoto, R. Parker, R. F. Smith, and S. M. Strain, *J. Chromatogr. Sci.* **10**, 708 (1972).

32. F. W. Karasek and P. W. Rasmussen, *Anal. Chem.* **44**, 1488 (1972).

SELECTED BIBLIOGRAPHY

Abbott, D., and R. S. Andrews, *An Introduction to Chromatography*, Houghton Mifflin, Boston, 1965.

Bobbitt, J. M., A. E. Schwarting, and R. J. Gritter, *Introduction to Chromatography*, Reinhold, New York, 1968.

Cassidy, H. G., in *Technique of Organic Chemistry*, Vol. X, A. Weissberger (ed.), Interscience, New York, 1957.

Giddings, J. C., and R. A. Keller (eds.), *Advances in Chromatography* 9 vols., Dekker, New York; continuing volumes to be published.

Heftman, E., *Chromatography*, 2nd ed., Reinhold, New York, 1967.

Lederer, E., and M. Lederer, *Chromatography*, Elsevier, New York, 1957.

Stock, R., and C. B. F. Rice, *Chromatographic Methods*, 2nd ed., Chapman and Hall, London, 1967.

GAS CHROMATOGRAPHY

Gas chromatography (GC) is unquestionably the most important laboratory separation technique. Its development since the original paper of Martin and James in 1952 (*1*) has sparked renewed interest in chromatography and the broader range of separation techniques. This chapter is presented before the one on liquid chromatography (LC) because its development preceded that of modern high-pressure LC.

Compared to gas–solid chromatography (GSC), gas–liquid chromatography (GLC) is far more popular because the range of applications is greater. The main application of GSC is the analysis of fixed gases. This chapter deals mainly with GLC.

THEORY

Basic chromatographic theory was presented in Chapter 11, but additional discussion is presented here relating to the extent of interaction of a solute with the stationary phase: the factors that determine the retention volume.

As a solute moves through a GLC column with the carrier gas (mobile phase), two forces cause it to dissolve in the stationary liquid phase. One is the equilibrium condition based on vapor pressure (defined by Raoult's law, Chapter 3), and the other is additional intermolecular forces (Chapter 5), which favor solubility in the liquid phase and cause the solute to deviate from Raoult's law. Thus the partial pressure of solute A, p_A, is

$$p_A = X_A \gamma_A p_A^0 \tag{1}$$

where X_A is the mole fraction of A in the solution, γ_A is the activity coefficient that corrects for deviations from ideal (Raoult's law) behavior, and p_A is the vapor pressure of pure A.

If two solutes, A and B, are introduced into a GLC column and do not interact, their respective elution (retention), times will reflect their respective interactions according to equation 1, and

$$\alpha = \frac{(V'_R)_A}{(V'_R)_B} = \frac{p_B^0 \gamma_B}{p_A^0 \gamma_A} \tag{2}$$

The ratio of their adjusted retention volumes is called the separation quotient α (Chapter 2); in GC it is also known as the *solvent efficiency* or simply the *relative retention.**

In most cases in GLC the activity coefficient provides the high degree of selectivity and makes GC more useful than distillation, which depends on vapor pressures alone. A classic example is the separation of benzene (b.p. 80.1°C), and cyclohexane (b.p. 81.4°C), which have nearly identical vapor pressures. They can be easily separated by GLC on a column containing dinonylphthalate because of their different activity coefficients (2):

$$\alpha = \frac{(V'_R)_{cyclohexane}}{(V'_R)_{benzene}} \approx \frac{\gamma_{benzene}}{\gamma_{cyclohexane}} \tag{3}$$

Activity coefficients can be calculated from GC data according to equation 4:

$$\gamma = \frac{1.7 \times 10^5}{V_g p^0 MW} \tag{4}$$

where MW is the molecular weight and V_g is the "specific" retention volume (the "net" retention volume at 0°C, per gram of liquid phase). In this example the activity coefficients at 326°K are 0.82 and 0.52 for cyclohexane and benzene, respectively (2), and $\alpha = 1.6$. The V_R for benzene is 1.6 times that of cyclohexane largely due to the attraction of its π-electrons with those of the stationary phase, dinonylphthalate.

SOME SYMBOLS AND DEFINITIONS

In the early days of GC it became evident that there was a need for standardization of terms and symbols, so various groups published recommended practices for reporting data. These include the IUPAC (3), ASTM (4), and the British Institute of Petroleum (5, 6). Unfortunately the recommendations are not uniform and not easily adapted to LC, so the choice of symbols for this book was difficult. The IUPAC has also attempted to standardize chromatographic symbols (7). Wherever possible, the officially recognized terms and symbols have been used. For GLC the subscripts G for gas and L for liquid replace the more general subscripts M for mobile and S for stationary.

* Note that the "prime" means "adjusted"; the ratio of unadjusted retention volumes is different.

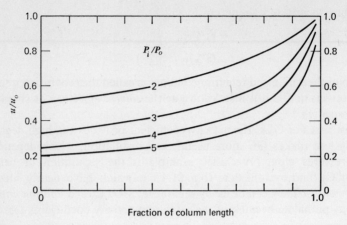

Figure 1. Effect of pressure on flow through column (P_i/P_0 = ratio of inlet to outlet pressure; u/u_0 = ratio of carrier velocity at a particular point in the column to the outlet velocity).

The one characteristic of GC that needs special attention is the compressibility of the mobile gas phase. The volume of the carrier gas is lower under the high pressure of the column inlet (P_i) than it is at the lower pressure (usually atmospheric) of the column outlet (P_0). The volumetric flow rate through the column is shown in Figure 1. As can be seen, the change in flow rate through the column is more drastic the higher the ratio of $\left(\dfrac{P_i}{P_0}\right)$. Note that the flow rate is greater at the column exit, where it is commonly measured. Consequently the measured volume is higher than the average volume in the column, and a correction factor is needed to get the average. The correction factor or so-called compressibility factor, j, is

$$j = \frac{3}{2}\left[\frac{(P_i/P_o)^2 - 1}{(P_i/P_o)^3 - 1}\right] \tag{5}$$

and the average pressure, flow rate, linear velocity, and volume are

$$\text{average pressure} = \bar{P} = \frac{P_0}{j} \tag{6}$$

$$\text{average flow} = \bar{F}_c = jF_c \tag{7}$$

$$\text{average linear velocity} = \bar{u} = ju \tag{8}$$

$$\text{corrected retention volume} = V_R^0 = jV_R = jF_c t_R \tag{9}$$

When applied to the retention volume, the term "corrected" refers specifically to the pressure correction and is designated by the superscript zero. Similarly the "corrected" gas volume (or dead volume) in the column, V_G^0, is

$$V_G^0 = jV_G \tag{10}$$

In summary, in GC the measured retention volumes have no superscripts (V_R), and the average or corrected volumes have a superscript zero (V_R^0, V_G^0). In LC no correction is necessary, but the measured retention volumes carry the superscript for consistency (V_R^0, V_L^0, V_M^0).

A closer consideration of the measurement of the gas flow rate reveals several other factors. First, the volume is usually measured at ambient temperature T_a, rather than the column temperature T_c. Second, the use of a soap film flow meter to measure the flow (Fig. 2) introduces water vapor into the measured volume, which necessitates a correction. Thus, if F_0 represents the measured outlet flow rate, the column flow rate at the outlet, F_c, is

$$F_c = F_0 \frac{T_c}{T_a}\left(\frac{P - p_w}{P}\right) \tag{11}$$

where the temperatures are in degrees Kelvin, P is atmospheric pressure, and p_w is the vapor pressure of water at T_a. The average or corrected flow rate is designated by $\bar{F}_c$. The symbol F with no subscript will be used to indicate the

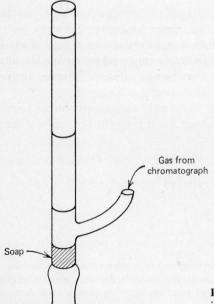

Gas from
chromatograph

Soap

Figure 2. Soap-bubble flowmeter for measuring outlet flow.

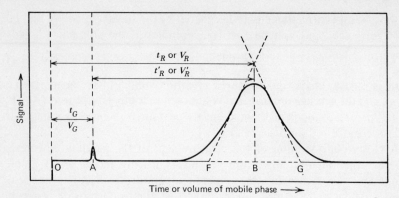

Figure 3. Typical chromatogram.

flow rate in those cases where the details just discussed are irrelevant or unimportant (such as in theoretical discussions and in LC).

Figure 3 shows a typical chromatogram for a single retained solute and a single nonretained solute (e.g., air). The latter serves to measure the volume of gas or "dead volume" in the column.* Since the x-axis can be given in units of time (at a constant flow rate) or volume, the dead volume is either t_G or V_G and is shown in Figure 3 as the distance O—A.

$$V_G = t_G F_c \tag{12}$$

Alternatively, the x-axis could be given as the "corrected" volume and the distance O—A would be equivalent to V_G^0. Since, for a given chromatogram, the flow rate F_c and the compressibility factor j are constant, any of these different parameters can be plotted; little attention is paid to the one actually being used if *relative* retention volumes are being considered. Hence, for the retained solute, the distance O—B can represent t_R, V_R, or V_R^0.

The other retention parameter, the "adjusted" retention volume, V_R', is also depicted in Figure 3. It is the distance A—B from the nonretained (air) peak to the retained peak. Thus

$$V_R' = V_R - V_G \tag{13}$$

If equation 13 is multiplied by j, the "corrected–adjusted" retention volume is V_N, the net retention volume:

$$jV_R' = V_N = V_R^0 - V_G^0 \tag{14}$$

The net retention volume is directly proportional to the partition coefficients,

* It has been assumed in all discussions in this book that there are no other sources of dead volume. In reality there is additional dead space in injection ports, detectors, and connecting tubing. Our discussions will be simplified if we continue to ignore these sources of dead volume.

as was shown in Chapter 11:

$$V_N = K_p V_L \tag{15}$$

A typical calculation of GC parameters is given in the Appendix.

INSTRUMENTATION

The heart of any chromatograph is the column, but most commercial instruments also include a sampling system and a detector. These were shown in Figure 8 of Chapter 11; commercial instruments are compared in reference 8.

A chromatograph can have up to four separately heated zones—one each for the column, the injection port, the detector, and the column–detector interface; the first three are most common. The column temperature may be increased during a run (programmed temperature), but the injection port and the detector are usually run isothermally.

Sample introduction is most often accomplished with a microsyringe through a self-sealing rubber septum. The injection port is designed so that the sample is swept quickly onto the column; it is usually heated to volatilize liquid or solid samples rapidly, or the samples are injected directly on the end of the column. The column temperature is usually set below the boiling point of the sample, and the injection port is set well above the boiling point. This leads to the common misconception that a liquid sample that is rapidly volatilized in the injection port is condensed when it reaches the cooler column. This is not the case; the column temperature is kept above the "dew point" of the sample, an operation facilitated by the very small sample size. A high percentage of the sample does dissolve in the stationary liquid (depending on its K_p or R_R), thus explaining the efficiency of on-column injection. If the column temperature is too low or the sample size too large, the partition isotherm will be nonlinear and the resulting peak will be "fronted" (see anti–Langmuir peak shapes in Chapter 5).

Figure 4 shows a simplified injection port design that keeps dead volume to a minimum and permits on-column injection. Other introduction systems use sample loops and valves.

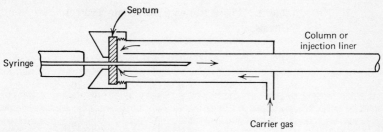

Figure 4. Simplified injection-port design.

Columns

Columns are usually made of metal (stainless steel, copper, or aluminum) or glass. The two types are packed and open tubular (OT). Packed columns vary in length from several feet to 50 ft, with 6 to 15 ft being most common. Outside diameters are usually 1/4 or 1/8 in. The columns are packed with a solid stationary phase (GSC) or inert solid support that contains the liquid stationary phase (GLC).

Open tubular (also known as capillary) columns were not used until 1957. In that year Golay introduced the idea based on an analogy with an equivalent electrical circuit. For many applications they are superior to packed columns; being "open" rather than "packed," they have a low pressure drop and a small sample capacity. Typical OT columns are approximately 30 to 500 ft long, with an inside diameter of 0.01 to 0.02 in. The stationary phase is coated on the inside wall about 0.5 μm thick (9); hence they are also known as wall-coated open tubular (WCOT) columns. Their β-values are large.

More recently an intermediate type of column has been used. A small amount of Celite solid support is coated on the inside wall of an open tube, resulting in a support-coated open tubular (SCOT) column, also known as a porous layer (PLOT) column (10). Like WCOT columns, they require special coating procedures (11). Some typical columns and β-values are given in Table 1.

Even though WCOT and SCOT columns have some definite advantages over packed columns, the latter are more commonly used, probably due to the high cost of OT columns ($100 and up), the difficulty in preparing them, and the special instrument requirements (e.g., sample splitters). Most of the subsequent discussion relates to packed columns. An adequate discussion of relative merits and characteristics of all three column types is beyond the scope of this introduction. Comparisons have been written by Ettre (12) and Guiochon (13); Ettre has also written a monograph on OT columns (14).

Table 1. Phase Ratios β for Some Typical Columns

Type	Diameter (in.)	β
Packed	1/8 O.D.	15–20
SCOT	0.02 I.D.	20–70
WCOT	0.01 I.D.	100–160
	0.02 I.D.	200–320

Solid Supports. In GLC a solid is required to hold the liquid phase immobile in the column. Ideally solid supports should be inert and have a large surface area. The most common types are derived from diatomaceous earth and sold under a variety of tradenames, of which Chromosorb (Johns–Manville Co.) is the most common (see Table 2). Chromosorb P has the largest surface area, but it is insufficiently inert for polar compounds; Chromosorb W is more inert but is quite friable; Chromosorb G is the newest support and is designed to have the inertness of W and the strength of P.

Table 2. Packings for GC

Name	Manufacturer[a]	Surface Area (m^2/g)	Packed Density (g/cm^3)	Pore Size (μm)	Maximum % Liquid Phase
Diatomaceous Earth:					
Chromosorb P	J	4.0	0.47	0.4–2	30
Chromosorb W	J	1.0	0.24	8–9	15
Chromosorb G	J	0.5	0.58	NA[b]	5
Chromosorb A (prep use)	J	2.7	0.48	NA	25
Fluorocarbon polymer:					
Teflon T-6	D	7.8	0.49	None	10
Glass					
Microbeads	—	0.01	1.4	None	3
Textured beads	C	0.04	1.35	NA	0.5

[a] Code: J = Johns–Manville; D = DuPont; C = Corning Glass (discontinued).
[b] NA = not available.

Other solids used as supports are fluorocarbon polymers, glass beads, and special silica products. These and the diatomites are listed in Table 2. Ottenstein has recently updated his paper on GC supports (*15*) with a review (*16*) that provides a good overview of the subject. It should be noted that the different densities of the solid supports require different percentages of liquid phase for equivalent liquid-phase loadings, as shown in Table 3 (*17*).

Diatomite supports are the most popular, but glass and fluorocarbon polymers are more inert. The undesirable surface activity of diatomites can be reduced by washing with acid and/or reacting the surface with a variety of "deactivators." The most common deactivating treatments are the following:

1. Acid washing (AW)
2. Methyl silicone (MS)
3. Hexamethyldisilizane (HMDS)
4. Dimethyldichlorosilane (DMCS)

Table 3. Equivalent Stationary Phase Loadings for Three Solid Supports[a,b]

Chromosorb P (wt %)	Chromosorb W (wt %)	Chromosorb G (wt %)
5.0	⎧ 9.3 ⎫	⎧ 4.1 ⎫
10.0	⎨ 12.9 ⎬	⎩ 8.3 ⎭
⎧ 15.0 ⎫	⎩ 25.7 ⎭	12.5
⎨ 20.0 ⎬	32.8	16.8
⎩ 25.0 ⎭	39.5	21.3
⎩ 30.0 ⎭	45.6	25.8

[a] From reference 17.
[b] Braces indicate the range recommended for each type.

The improvement in chromatographic performance is often phenomenal (16). The dimethyldichlorosilane treatment is usually the best.

For highest efficiency the solid support should have a small diameter and a narrow range of sizes. Table 4 lists some typical mesh sizes and actual particle diameters. In GLC a common mesh range is 80/100.

Stationary Liquids. Several hundred different liquids have been used as the stationary phases in GLC—so many, in fact, that the standardization of procedures is very difficult and the selection of a phase is complicated. A typical GC supplier lists just under 200 liquids but recommends that 21 of them will probably be adequate for most separations. Groups of chromatographers have made repeated attempts to standardize on a few liquid phases, but have been unsuccessful (18).

Table 4. Mesh and Particle Sizes

	Particle Diameter (μm)		
Mesh Range	From	To	Range (μm)
45/60	354	250	104
60/80	250	177	73
80/100	177	149	28
100/120	149	125	24
325/400	44	37	7

The main criterion in choosing a liquid phase is its selectivity toward the sample. Chapter 5 presented a general introduction to the nature of intermolecular forces, but the theoretical approach is inadequate for phase selection in GLC. An empirical method is given later in this chapter.

In practice, the liquid phase is often selected because of its low vapor pressure. It must not volatilize ("bleed") appreciably at the operating temperature, so high-boiling liquids and polymers are commonly used. Some are listed in Table 5 along with their upper and lower temperature limits. This

Table 5. Maximum and Minimum Temperatures for Some Common Liquid Phases

Phase	Temperature (°)	
	Maximum	Minimum
Dinonylphthalate	150	20
Apiezon L	250	50
DC-200 Silicone (50 cSt)	250	−90
DC-200 (500 cSt)	250	−60
DC-200 (12,500 cSt)	250	0
Carbowax 20M	250	60
OV-1	350	100
Dexsil 300	500	50

minimum temperature is the freezing point of the liquid or the point at which its viscosity becomes so great that its solvent characteristics change (19). This can be seen by comparing the minimum temperatures for a range of silicone oils ranging from 50 to 12,500 cSt viscosity; the lower the viscosity, the lower the minimum temperature.

A new type of stationary liquid phase is made by chemically bonding the liquid to the solid support. As noted earlier, the common solid supports have active sites on their surfaces that can be chemically deactivated. Therefore it is not unexpected that they can react with a variety of chemicals to form useful phases. Two types of chemically bonded phases are available commercially at present: Durapak from Waters and Permaphase from DuPont. The former

has an—Si—O—R or ester linkage, and R can be either *n*-octane, Carbowax

400, or oxydipropionitrile. Information about the Permaphase materials has

not been published, but they probably involve —Si—R bonds where R is an ether, a nitrile, or a hydrocarbon (20). These materials are also used in LC, and other types of bonded phase are being investigated. Since the bonded phase is present essentially as a monolayer, the question can be raised as to whether the stationary phase is a liquid or a solid. Further discussion of these phases is found in Chapter 13.

The amount of liquid phase on a conventional type packing varies from less than 1 to 30%, with most falling between 3 and 10%. According to the rate equation, the lower levels are preferable since they produce a small film thickness.

Finally, it should be noted that the maximum sample size that can be used depends on the amount of liquid phase. When too large a sample is used, the column becomes overloaded, K_p is no longer constant (partition isotherm departs from linearity), and peaks exhibit asymmetry with leading edges. For example, in a 1/4-in. packed column the maximum sample size should be about 0.5 to 1 mg per peak; but in a 0.01-in. WCOT column, samples should not exceed 5 μg per peak.

Active Solids.   The most common types of solid used in GSC are listed in Table 6. They are used mainly in the analysis of fixed gases and light hydrocarbons (21). Molecular sieves are particularly useful for the separation of oxygen and nitrogen, and bonded phases have already been mentioned.

Table 6.   Active Solids for GSC

Type (Trade Names)	Uses and Special Properties
Carbon (Carbosieve-B[a])	Fixed gases; light hydrocarbons O_2–N_2; water elutes before methane
Silica (Spherosil, Porasil)	Fixed gases; light hydrocarbons
Alumina	Hydrocarbons
Molecular sieves	H_2–O_2–N_2; n- and branched-chain hydrocarbons
Bonded phases (Durapak, Permaphase)	Reduced column bleed (compared to GLC)
Porous polymers (Porapak, Chromosorb Century Series)	Aqueous solutions, acids, and other polar materials
Organoclays plus stationary liquid (Bentone 34)	Aromatic hydrocarbon isomers (23)

[a] Can also be classified as a molecular sieve.

Porous polymers used in GSC are copolymers of styrene and a vinylbenzene (*22*). In principle these polymers should not bleed, but most of them do until the lower-molecular-weight oligomers are removed; they decompose at temperatures higher than about 250°C.

Carrier Gas

The choice of a carrier gas is usually determined by detector characteristics and cost. The popular thermal conductivity detector requires hydrogen or helium for good sensitivity, whereas the flame ionization detector can be operated with the less expensive nitrogen. For traditional GC, the carrier gas should be inert and have no effect on the chromatographic process other than to provide a mobile phase. It must be very pure.

Detectors (*24*)

Virtually every conceivable means of detecting gases and vapors has been exploited in designing GC detectors, and over 100 different ones have been described. However, only a few detectors have become popular and are commercially available. It is estimated that over 90% of the GC work is done with the thermal conductivity detector (TCD) or the flame ionization detector (FID). Consequently they receive the most attention in this chapter.

Thermal Conductivity. The TCD consists of a metal block containing cavities through which the gas flows. A typical design is shown in Figure 5. There must be a minimum of two cavities—one for the sample and the other for the reference—but most detectors have four cavities (two for each) and twice the sensitivity. A transducer is located in each cavity and can be a thermistor or a resistance wire (e.g., tungsten). The transducers are heated above the temperature of the metal block and lose heat to it at a rate depending on the thermal conductivity of the gas flowing through the cavity. If, as is commonly the case, the carrier gas has a high thermal conductivity (e.g., helium), the transducer runs at a relatively low temperature. When a sample elutes from the column and enters the sample side of the TCD, the thermal conductivity of the mixture goes down, less heat is transferred away from the transducer, and its temperature goes up. This causes a resistance change in the transducer: a decrease for thermistors and an increase for resistance wires. Because of their respective temperature coefficients of resistance, thermistors are often used below 100°C and "hot wires" above 100°C. Hot wires are useful over a wider temperature range and are more popular than thermistors.

Top view

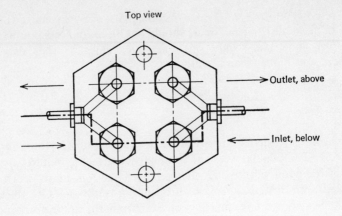

Outlet, above

Inlet, below

Side view

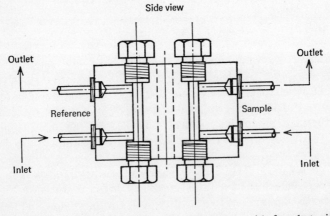

Figure 5. Typical thermal conductivity detector. Shown with four hot wires mounted concentrically. Courtesy of the Gow–Mac Instrument Co.

The change in resistance is converted to an electrical signal by incorporating the transducers in a Wheatstone bridge circuit like that shown in Figure 6. The bridge is balanced (using the "zero" control) when pure carrier gas is flowing through both parts of the cell; the bridge is unbalanced when a sample appears in the cell and a voltage develops across X–Y. This voltage is fed to a voltage divider (attenuator) and then to a recorder, and becomes the signal. Amplification is not usually necessary, and the bridge operates "out of balance" rather than at "null" (which is the more common mode, for non-chromatographic applications).

Table 7 lists some relative thermal conductivities. It can be seen that maximum sensitivity will be obtained for most organic compounds if a carrier

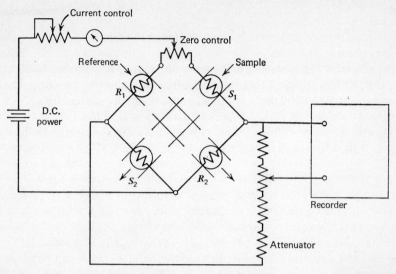

Figure 6. Simplified Wheatstone bridge circuit for a thermal conductivity detector. Four elements: R_1 and R_2 for reference and S_1 and S_2 for sample.

Table 7. Thermal Conductivities and TCD Response Values for Selected Compounds

Compound	Thermal Conductivity[a]	RMR[b]
Carrier gases:		
Argon	12.5	—
Carbon dioxide	12.7	—
Helium	100.0	—
Hydrogen	128.0	—
Nitrogen	18.0	—
Samples		
Ethane	17.5	51
n-Butane	13.5	85
n-Nonane	10.8	177
i-Butane	14.0	82
Cyclohexane	10.1	114
Benzene	9.9	100
Acetone	9.6	86
Ethanol	12.7	72
Chloroform	6.0	108
Methyl iodide	4.6	96
Ethyl acetate	9.9	111

[a] Relative to He = 100.
[b] Relative molar response in helium. Standard: benzene = 100.

gas of high thermal conductivity is used. However, from the relative response values given in the table, it can also be seen that response is not directly proportional to the difference in thermal conductivity between the carrier and the sample; this indicates that the mechanism by which the detector operates is more complex than just described. Further details can be found in GC textbooks and in the review papers of Lawson and Miller (*25*), Richmond (*26*), and Johns and Stepp (*27*). Although the TCD is only moderately sensitive, it is universal in response, rugged, simple, and inexpensive.

Flame Ionization Detector. The FID is a small oxy-hydrogen flame in which the sample is burned. A typical design is shown in Figure 7. Hydrogen is mixed with the carrier gas (and sample) before entering the burner, and air (or oxygen) is caused to flow alongside the burner. A high voltage (about 300 V DC) is impressed across electrodes in the vicinity of the flame so that the ions and electrons formed in the combustion of a sample can be collected to form a current that becomes the signal. This current is very small (10^{-12} to 10^{-9} A) and must be amplified before it can be recorded. A very small current

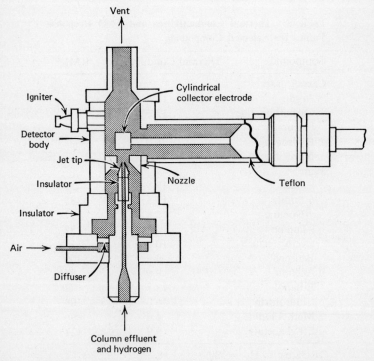

Figure 7. Schematic of a flame ionization detector. The detector body and jet tip are at the same potential. (Courtesy of the Perkin–Elmer Corp.)

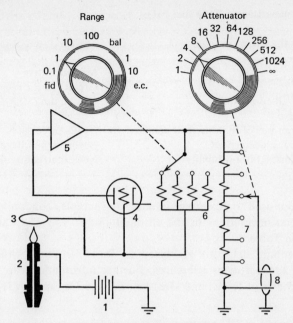

Figure 8. Schematic circuit for a flame ionization detector. Key: 1, detector polarizing voltage; 2, flame tip; 3, collecting anode; 4, electrometer tube; 5, operational amplifier; 6, range (feedback resistor); 7, output attenuator; 8, recorder terminals.

flows when no sample is being burned due to impurities and leakage, so a bucking voltage (zero) is also necessary. A simplified circuit is shown in Figure 8. Further details can be obtained from textbooks and the recent review of Bocek and Janak (*28*), which contains over 300 references, and that of Blades (*29*).

The main advantage of the FID is its high sensitivity, but since it requires two additional gases and a high-impedance amplifier, it is more complex than the TCD. Some compounds are not detected (Table 8), including water,

Table 8. Some Compounds Not Detected by the Flame Ionization Detector

He	O_2	NO	CS_2
Ar	N_2	NO_2	COS
Kr	CO	N_2O	$SiCl_4$
Ne	CO_2	NH_3	$SiHCl_3$
Xe	H_2O	SO_2	SiF_4

Table 9. Comparison of TCD and FID

Thermal Conductivity Detector	Flame Ionization Detector
Simple and troublefree	More sensitive
Lower cost	Greater linear range
Responds to all compounds	Can be used with OT columns
Nondestructive	
Needs a carrier gas of high thermal conductivity	Needs three gases
Needs stable low-voltage DC	Needs high-voltage DC
Needs good temperature stability	Needs current amplifier

which is often used to advantage in analyzing samples containing water that would otherwise interfere in the chromatogram. The TCD and FID are compared in Table 9.

Other common detectors are listed in Table 10 along with their main advantages and pertinent references. Further information can be obtained from the reviews of Krejci and Dressler (*30*) and Hartmann (*31*).

Table 10. Some Special Detectors

Detector	Major Advantage or Use	Approximate Detection Limit (g/sec)	Reference
Gas density	Quantitative analysis; corrosives	10^{-10}	*32*
Electron capture	Halogens, conjugated	10^{-14}	*33*
Alkali flame (thermionic)	Phosphorus	10^{-13}	*34*
Flame photometric	Phosphorus and sulfur	10^{-12}	*35*
Microcoulometric	Nitrogen, sulfur, phosphorus, halogens	10^{-10}	*36*

RETENTION PARAMETERS AND QUALITATIVE ANALYSIS

From the preceding section it should be clear that the retention volume for a given solute under a given set of conditions (temperature, liquid phase, column, etc.) is a constant, just as the partition coefficient is a constant. Hence the quantity V_N is characteristic of a given substance and can be used to identify it. Limitations arise from the inability to reproduce a given set of constant conditions from one laboratory to another, or indeed from one day to the next, as well as from a recognition that several substances will likely have exactly the same V_N. However, an unknown can be identified from among a limited number of possibilities simply by adding a small amount of the suspected unknown to the sample and rechromatographing it, noting if the area (height) of the suspected peak has increased.

Alternatively, retention volumes can be made more reliable by measuring them relative to a standard:

$$\text{relative retention} = \frac{(V_N)_A}{(V_N)_B} = \frac{(V'_R)_A}{(V'_R)_B} = \frac{(K_p)_A}{(K_p)_B} \qquad (16)$$

Benzene is sometimes used as the standard, but there is no universal standard that permits interlaboratory comparisons. We shall return to this problem shortly.

First, it is necessary to note the retention behavior of members of a homologous series on a given column at a given temperature. The retention volumes increase logarithmically, as shown in Figure 9. If log V_N is plotted versus the carbon number, as shown in Figure 10, a straight line results.* This is the relationship we expect based on the discussions of thermodynamics in Chapter 3. We saw that

$$\log K_p = -\frac{\Delta \mathscr{H}}{2.3 \mathscr{R} T} + \text{constant} \qquad (17)$$

and that $\Delta \mathscr{H}$ is approximately proportional to the boiling point of a liquid:

$$\Delta \mathscr{H} = 21 T_{\text{boil}} \qquad (18)$$

Furthermore, the boiling points of the members of a homologous series increase with the carbon number in a regular fashion, so $\Delta \mathscr{H}$ as well as K_p

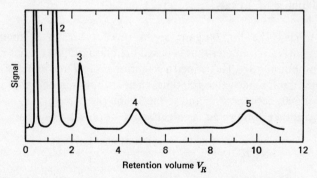

Figure 9. Typical chromatogram for a homologous series of compounds.

* The retention parameter V_N is used in these discussions for the sake of theory. However, it should be noted that, experimentally, the flow rate and pressure (and hence j) are constant, so it is simpler and equally correct to use t'_R, the adjusted retention time.

or log V_N should increase in proportion to the number of carbons, as shown in Figure 10. Use is made of this relationship in qualitative analysis and in predicting the retention volumes of higher members of a homologous series.

The reliability of an identification can be improved if the sample is run on two different columns rather than on only one. Homologous-series plots based on data from two columns are shown in Figure 11a, which is linear, and Figure 11b, which is logarithmic. Different homologous series lie on different straight lines in both cases, but the log–log plot has the advantage that the lines do not converge to the origin.

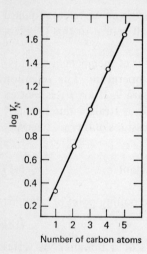

Figure 10. Plot of carbon number versus log net retention volume for a homologous series.

Number of carbon atoms

Retention Index of Kovats

Kovats utilized the homologous series relationship in recommending a relative-retention parameter that is based on a homologous series rather than a single compound (37). He defined a retention index I in which the n-paraffins are assigned a reference value 100 times their carbon number; thus hexane has an index of 600, heptane 700, and so on. Using the paraffins as standards, the index for another compound is calculated as follows:

$$I = 100\left[\frac{(\log V_N)_u - (\log V_N)_x}{(\log V_N)_{x+1} - (\log V_N)_x}\right] + 100x \qquad (19)$$

where u stands for unknown and x for the paraffin with x carbons and eluting just before the unknown; $x + 1$ stands for the paraffin eluting just after the unknown and having $x + 1$ carbons. This is shown graphically in Figure 12, which is the easiest method of arriving at index values experimentally. The

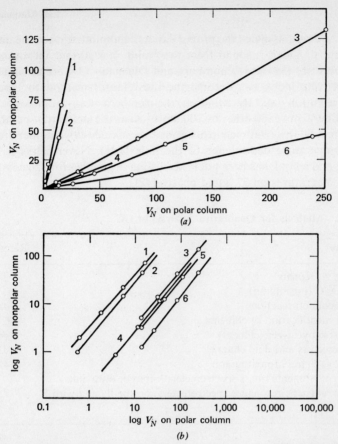

Figure 11. Homologous series plots: (*a*) linear; (*b*) logarithmic (1, alkanes; 2, cycloalkanes; 3, esters; 4, aldehydes; 5, ketones; 6, alcohols).

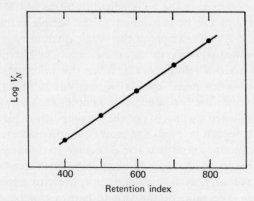

Figure 12. Retention index (Kovats) plot.

paraffins are run, assigned the proper indices, and plotted; then the unknown is run and its index is found from the graph. For further information the papers by Ettre (*38*) and Schomburg and Dielmann (*39*) are recommended.

The Kovats index has become the most useful method for specifying relative-retention data. By definition, the members of any homologous series should differ from each other by 100 units just as the standard *n*-paraffins do. Unfortunately, this relationship does not always hold (*40*). Another drawback is the temperature dependence of the index (*41*). Nevertheless it is very popular, and a book has been published listing the indices for 350 compounds at two temperatures on 77 liquid phases (*42*).

Table 11. Methods for Qualitative Analysis in GC

Technique	Reference
Indices (e.g., Kovats)	*37*
Pyrolysis GC (precolumn)	*37, 44*
Other precolumn reactions	*43*
Chemical identification of effluents	*43, 45*
Vapor-phase pyrolyzer (effluent)	*46*
Special detectors and dual channel	*43*
Molecular-weight chromatograph	*47*
Auxiliary instruments (mass spectrometer; infrared, ultraviolet, nuclear magnetic resonance spectrometers) on-the-fly and with trapping	*43*

In addition to the Kovats index and other retention parameters, a wide variety of methods are used for qualitative analysis. They are listed in Table 11, and most are covered in the book edited by Ettre and McFadden (*43*).

Chemical reactions can be performed on the sample before or after it is chromatographed. As an example of the latter, qualitative group tests have been made on individual peaks as they elute simply by bubbling the column effluent through a set of solutions (*45*). When this information is combined with homologous-series plots, qualitative analysis is facilitated. Pyrolysis chromatography finds use for nonvolatile samples such as polymers, which are pyrolyzed, followed by analysis of the decomposition products by GC.

Selective detectors can be used. The most common ones are listed in Table 10. To make maximum use of selective detectors, two different ones can be connected in parallel at the end of a column. Both detector signals are recorded simultaneously, each responding selectively to a special element or functional group. The ratio of the two responses can be correlated with compound type for qualitative analysis.

The molecular-weight chromatograph is a special instrument using two gas-density balances and two different carrier gases. It can be used to determine the molecular weight of the components of a sample in the range from 2 to over 400 (47).

Finally, it is possible to use another instrument for identifying an eluted peak. In one case, the sample eluting from the column is trapped (e.g., in a cooled cell) and taken to the other instrument for analysis. But, if the auxiliary instrument can perform a rapid analysis (during the time of solute elution), it can be attached directly to the GC column outlet. Two powerful techniques for identification, infrared (IR) and mass spectrometry (MS), have been adapted for use in either mode. Many other methods have been used, including nuclear magnetic resonance, ultraviolet absorption, Raman, fluorescence, and TLC. The coupling of a mass spectrometer to the exit of a gas chromatograph usually requires the use of an interface to handle the pressure differential (48).

CLASSIFICATION OF LIQUID PHASES

Some of the desirable characteristics of stationary liquid phases were discussed earlier in this chapter, but the important topics of liquid-phase selectivity was deferred until after the qualitative analysis section. The reasons will become obvious.

Background material for this section includes the discussion on thermodynamic principles in Chapter 3 and the description of intermolecular forces in Chapter 5. However, the theoretical approach will not be pursued further since no theory adequately covers the selection of a liquid phase and good reviews of the present situation are available (49, 50). Instead we turn our attention to empirical methods.

The basic premise in the empirical methods is that a given retention behavior and hence a given solute–solvent interaction can be attributed partly to the solute and partly to the solvent. In order to assign numerical values to represent the magnitude of these interactions, some reference standards must be arbitrarily chosen and a scale must be set up. This can be effected by choosing reference standards at both extremes of the "polarity" scale or by choosing an existing scale. The final choice is the number of standards to be used. Obviously a larger number of standards gives better results but requires more time and is more complex.

Rohrschneider has been very active in this work. One of his early suggestions was a "polarity scale" that assigned a value of zero to the hydrocarbon squalane ($C_{30}H_{62}$) and 100 to β,β'-oxydipropionitrile (ODPN) (51). Butane and butadiene were chosen as the two solutes to be chromatogrammed on the liquid phases to be characterized. Although this approach was successful it was soon replaced by a more elaborate system.

Rohrschneider Constants

A more elaborate system proposed by Rohrschneider is based on the Kovats index scale and is the one usually associated with his name. His original paper is in German (52), but a summary of its usage is available in English (53). The

Table 12. Compounds Used for Liquid-Phase Characterization

	Solute Used by	
Designation	Rohrschneider	McReynolds
a	Benzene	Benzene
b	Ethanol	n-Butanol
c	2-Pentanone (MEK)	2-Pentanone
d	Nitromethane	Nitropropane
e	Pyridine	Pyridine
		2-Methyl-2-pentanol
		Iodobutane
		2-Octyne
		1,4-Dioxane
		cis-Hydrindane

system has become quite popular. Squalane remains as the one reference standard of zero polarity.

Five solutes are used to characterize a liquid phase; each is intended to measure a different type of solute–solvent interaction (see Table 12). For example, benzene measures dispersion forces and π-bonding, whereas ethanol measures hydrogen bonding (as an acceptor and a donor). Each solute is run on squalane and the liquid phase of interest at 100°C and 20% liquid loading. The Kovats index is determined for each solute, and the difference (ΔI) between the two values on the two liquid phases is obtained. The ΔI for all the solutes is given as

$$\Delta I = ax + by + cz + du + es \qquad (20)$$

where a, b, c, d, and e represent the five solutes listed in Table 12 and x, y, z, u, and s represent the five constants that characterize the liquid phase. Since a

represents benzene, x is a measure of the dispersion and π-bonding forces of the liquid phase, and so on. Note that a has a value of 100 for benzene and a value of zero for all other solutes. Likewise b is 100 for ethanol and zero for all others. Technically, then, each of the x, y, z values should equal ΔI divided by 100 as reported by Supina and Rose (53), and the ΔI in equation 20 is actually the sum of five ΔI values. Before examining some typical values for some liquid phases, another development must be considered.

Table 13. Rohrschneider–Type Constants for the Characterization of Liquid Phases[a]

Liquid Phase	x'	y'	z'	u'	s'	m	i	o	d	h
Squalane	0	0	0	0	0	0	0	0	0	0
Apiezon L	35	28	19	37	47	16	36	11	33	33
SE-30	15	53	44	64	31	31	3	22	44	-2
Di-n-octyl phthalate	96	188	150	236	172	144	92	68	142	27
Di-i-octyl phthalate	94	193	154	243	174	149	92	69	147	24
Didecyl phthalate	136	255	213	320	235	201	126	101	202	38
Tricresyl phosphate	176	321	250	374	299	242	169	131	254	76
QF-1	144	233	355	463	305	203	136	53	280	59
OV-210	146	238	358	468	310	206	139	56	283	60
Triton X-100	203	399	268	402	362	290	181	145	304	83
XE-60	204	381	340	493	367	289	203	120	327	94
Carbowax 20M	322	536	368	572	510	387	282	221	434	148
Diethylene glycol succinate	492	733	581	833	791	579	418	321	705	237
1,2,3-Tris(2-cyanoethoxy) propane	594	857	759	1031	917	680	509	379	854	269

[a] From McReynolds (54).

In 1970 McReynolds published similar constants for over 200 liquid phases (54). However, he made two changes in his selection of standard solutes: first, he used some higher homologs (butanol instead of ethanol) and, second, he extended the list to 10 compounds. His solutes are also given in Table 12. Also, he did not divide his values by 100. McReynolds's publication provides the largest listing of Rohrschneider–type constants, and some are given in Table 13. All 10 values are given for each liquid phase, but McReynolds has indicated that the first seven are the most useful in characterization, so it is highly probable that future studies will be limited to seven. In fact, one supplier of liquid phases lists six of the constants for most of the liquid phases in his catalog.

Table 14.   Rohrschneider Constants for Some Solutes[a]

Solute	a	b	c	d	e
Benzene	100	0	0	0	0
Ethanol	0	100	0	0	0
2-Pentanone	0	0	100	0	0
Nitromethane	0	0	0	100	0
Pyridine	0	0	0	0	100
i-Propanol	−18.2	+95.9	+15.8	−6.5	+2.1
n-Propanol	−9.4	+105.3	+0.3	+6.6	−7.5
Acetone	−5.3	−4.6	+94.9	+7.9	+5.6
n-Butyl acetate	−3.8	−13.3	+57.3	+13.9	+20.0
Propionaldehyde	+13.3	−1.0	+74.9	+4.8	+1.3
Di-n-butylether	+17.3	+9.8	+29.7	−12.5	−2.8
Cyclohexane	+32.1	−22.5	−21.6	+4.1	+29.7
Chloroform	+69.7	+28.9	−72.6	+53.1	−6.3
Toluene	+108.3	+3.8	+8.8	−7.0	−7.6

[a] From Rohrschreider (*52*).

With this information it has been possible to find liquid phases that have nearly identical values and retention characteristics. This has been particularly useful in those cases where a preferred liquid phase is no longer available and as an aid in reducing the number of liquid phases needed in the laboratory. Note, for example, that the new liquid phase OV-210 is nearly identical with the obsolete QF-1.

The order of listing in Table 13 is according to increased "average polarity" based on the first five constants. Thus, in any column, deviation from a regular increase in value represents an unusually high or low interaction. Tricresyl phosphate, for example, has high values for x' and y' representing its interactions with an aromatic compound (benzene) and an alcohol (butanol), so it would be expected to cause longer retention times for those two types of compound. Another example is QF-1 (and the equivalent OV-210), which has a high z' value, representing a high interaction with ketones, and a very low o-value, indicating a lack of interaction with alkynes. Such special selectivities can be used in choosing a liquid phase to solve a specific separation problem.

This system can be taken one step further by using the x, y, and z constants for the liquid phases in the determination of the a, b, and c constants for other solutes. The procedure is to run the new solute on five* different liquid phases

* We are referring back to the original work of Rohrschneider (*51*), which used only five terms.

whose constants are known. The five ΔI values obtained are substituted into five equations (like equation 20) and they are solved simultaneously for the five unknown a, b, c, d, e values. Rohrschneider (52) has reported some results, which are included in Table 14. As expected, the c value for acetone, a ketone, is nearly 100, and the other four values are close to zero, as is the case with the "standard" ketone, 2-pentanone. Similarly, toluene has an a value of nearly 100, and the others are close to zero.

In principle, one can determine the Rohrschneider constants for the solutes in his samples and then calculate the Kovats index for each solute on any liquid phase available using equation 20. Thus the best liquid phase to perform a given separation can be selected. Obviously this becomes quite complex if the sample is complex and impossible if the components of the sample are unknown. Still, the empirical Rohrschneider system represents the best method known for selecting a liquid phase.

Very Selective Liquid Phases

A few liquid phases are highly selective and are used for specific analyses. Table 15 summarizes the main types and provides references in addition to the summary by Karger (49).

The silver salts are used with a conventional liquid phase (e.g., a glycol) and are thought to form complexes with olefins that are found to be strongly retained. Bentone 34 and the newer liquid crystals operate in part on the ability of a solute molecule to fit into the stationary phase lattice, and thus the retention is influenced by the size and shape of the solute.

Two approaches have been used to separate optical isomers. In one case the racemates are first converted to diasterioisomers, which requires a careful choice of resolving agent and is not too dependent on the stationary phase. In the other case the racemates are separated directly, which requires a very selective stationary phase, for example, an optically active one.

Table 15. Very Selective Liquid Phases

Type	Example	Reference
Complexes	Silver salts	55
Shape	Bentone 34	23
	Liquid crystals	56
Optical activity	—	57

TEMPERATURE EFFECTS

The effect of temperature in GC is very important but also very complex. Most important is the effect on the partition coefficient according to the Clausius–Clapeyron equation, which has already been discussed. Since, for a given column and temperature, K_p is proportional to k' and to V_N, any of these parameters can be used to discuss temperature dependence. Typical is Figure 13, which is a plot of $1/T$ versus log V_N. The slope of the lines is proportional to $\Delta\mathcal{H}$, so the lines would be parallel if $\Delta\mathcal{H}$ were the same for all compounds shown.

This is true as a first approximation, but close inspection reveals that the curves diverge at lower temperatures. Thus, as a general rule, the best

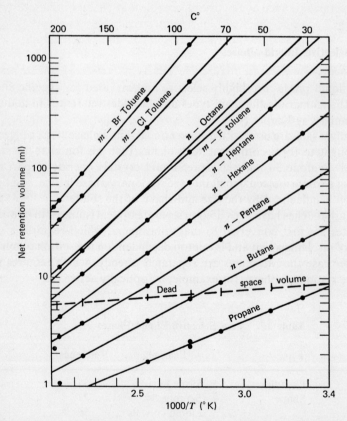

Figure 13. Temperature dependence of retention volume. Reprinted with permission from W. E. Harris and H. W. Habgood, *Talanta* **11**, 115 (1964), Pergamon Press.

separations are achieved at the lowest possible temperatures. It is even possible that the lines for two compounds will cross (e.g., n-octane and m-fluorotoluene in Figure 13) so that the order of elution is reversed as the temperatue is changed, and the compounds cannot be separated at all at one temperature (the crossover point).

Temperature also affects the viscosity of the carrier gas, and hence the pressure drop and flow rate. However, regulators are available that will maintain a constant flow rate, thus negating this effect. Finally, there is the effect of temperature on column efficiency, n or H, which is quite complex and has been discussed by Harris and Habgood (58). No generalizations can be drawn, but the overall effect is dependent on the relative temperature dependence of the following terms in the van Deemter equation: D_M, k', j, D_S, and D_L. Usually the effect on n is minor and is of considerably less importance than the effect of temperature changes on retention time.

Temperature is such an important variable that an important technique has been developed whereby the column temperature is increased with time during the analysis of a sample. This is called temperature programming.

Programmed-Temperature GC. The effect of increasing the temperature during an analysis can best be seen from a typical analysis like that shown in Figure 14. In Figure 14b, the programmed-temperature run, the separation of the early peaks is good and the total time for the analysis is optimized. When this sample is run isothermally, one has the choice of either of these two benefits, but not both. Note that the peak widths in PTGC are nearly identical rather than increasing with increasing retention time.

A simple but adequate treatment of PTGC has been given by Giddings (59). Some of the main points will be summarized, but for a complete treatment the monographs by Harris and Habgood (60) and Mikkelsen (61) can be consulted.

We saw in Figure 13 how the retention volume depends on the temperature. Let us now determine approximately what temperature increase is required to cut the retention volume (or partition coefficient) in half:

$$\frac{(K_p)_1}{(K_p)_2} = 2 = \frac{\exp\left(-\Delta\mathcal{H}/\mathcal{R}T_1\right)}{\exp\left(-\Delta\mathcal{H}/\mathcal{R}T_2\right)} = \exp\left[\frac{\Delta\mathcal{H}}{\mathcal{R}T}\left(\frac{\Delta T}{T}\right)\right] \tag{21}$$

where ΔT is the average of T_1 and T_2. Taking the log and rearranging, we get

$$\Delta T = \frac{0.693\,\mathcal{R}T^2}{\Delta\mathcal{H}} \tag{22}$$

Assuming Trouton's rule that $\Delta\mathcal{H}/T_b = 21$ and a boiling temperature of a

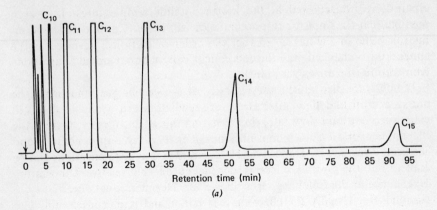

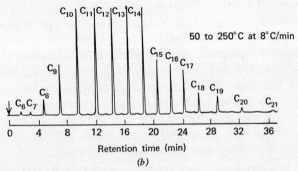

Figure 14. A comparison of isothermal (*a*) and temperature-programmed (*b*) gas chromatography. Sample of *n*-paraffins. Reprinted by permission of Varian Associates, from *Basic Gas Chromatography* by McNair and Bonnelli.

typical sample of 227°C (500°K),

$$\Delta T = \frac{0.693 \times 2 \times (500)^2}{21(500)} \approx 30°C \tag{23}$$

Thus we can say that an increase in 30°C will cut the retention volume in half.

The effect temperature programming has on a solute's migration through a column is depicted in Figure 15 using the 30°C value just derived. The step function is based on the assumption that R_R will double every 30°C, but the smooth curve drawn through steps is closer to reality, of course. If the carrier-gas velocity u is constant, the rate of the solute's movement through the column will be proportional to R_R since its velocity is ($R_R \times u$). If l is taken as the distance the solute moved through the column in the last 30°C rise in temperature, then $l/2$ is the distance it moved in the previous (next to last) 30°C, $l/4$ in the 30°C before that, and so on. The sum of these fractions approaches $2l$, which is the total length L. Hence the solute traveled through the last half of the column in the last 30°C increment. This is quite evident from the figure, and it is an interesting and useful concept in understanding how PTGC works. In effect, the components of a mixture remain near the entrance of the column and "boil off," moving down the column as the temperature is raised.

One means for relating isothermal performance with programmed temperature is to define a single weighted-average temperature for the programmed run that can serve to characterize it. Giddings has called this the *significant temperature*, T' and has shown its relationship to the final programmed temperature T_f. It is already obvious that T' will be close to T_f since most of the solute movement occurs in the last 30°C.

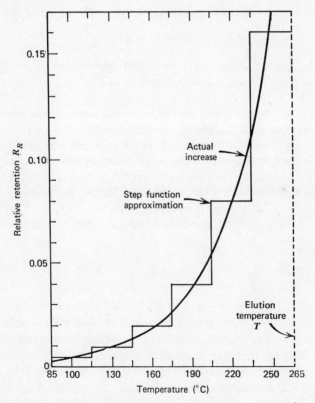

Figure 15. Step-function approximation to programmed-temperature gas chromatography. Reprinted from J. C. Giddings, *J. Chem. Educ.* **39**, 569 (1962), by courtesy of the Chemical Education Division, American Chemical Society.

The average temperature of the last temperature step (Fig. 15) is $(T_f - 15)$. Since half the solute migration occurs during this temperature zone, the weighted average of the total solute migration time is

$$\tfrac{1}{2}(T_f - 15) \tag{24}$$

Similarly, for the previous zone the average temperature is $(T_f - 45)$ and the corresponding fraction is $\tfrac{1}{4}$. Thus its weighted average is

$$\tfrac{1}{4}(T_f - 45) \tag{25}$$

Continuing in this fashion, we obtain

$$T' = (T_f - 15)\tfrac{1}{2} + (T_f - 45)\tfrac{1}{4} + (T_f - 75)\tfrac{1}{8} + \cdots \tag{26}$$

$$= T_f - 15 - 30(\tfrac{1}{4} + \tfrac{2}{8} + \tfrac{3}{16} + \cdots) \tag{27}$$

The sum of the numbers in parentheses approaches unity as a limit, so

$$T' = T_f - 45 \tag{28}$$

We can infer, then, that the best isothermal temperature should be the same as T', and our programmed run would have as its final temperature one that is 45°C higher.

The effect that programming the temperature has on retention was shown in Figure 14. Specifically, the spacing between members of a homologous series is linear rather than logarithmic, and the peak shapes and widths are nearly constant. As a consequence of the latter phenomenon, calculations of the number of theoretical plates are not valid for comparisons with isothermal runs unless a new definition for n is used for PTGC.

An annotated bibliography of PTGC covering 1952 to 1964 has been published in three parts by Harris and Habgood (62).

OPTIMIZATION

A discussion on the optimization of the performance of a GC column must begin with a consideration of the measures of column performance. Three of these measures are resolution, number of theoretical plates, and plate height. If we can assume that our problem is the separation of A from B then resolution is a better measure than the other two, although they are related.

In Chapter 2, resolution R_s was defined as

$$R_s = \frac{2d}{w_A + w_B} \approx \frac{d}{w} \tag{29}$$

Since $d = [(t_R)_B - (t_R)_A]$ and $w = 4\sigma$, R_s can also be written as

$$R_s = \frac{(t_R)_B - (t_R)_A}{4\sigma} \tag{30}$$

Furthermore, since

$$n = \left(\frac{t_R}{\sigma}\right)^2 \tag{31}$$

then

$$R_s = \frac{1}{4} n^{1/2} \left[1 - \frac{(t_R)_A}{(t_R)_B}\right] \tag{32}$$

We have also seen (Chapter 11) that

$$R_R = \frac{t_M}{t_R} \tag{33}$$

and (Chapter 2)

$$R_R = \frac{1}{1 + k'} \tag{34}$$

Making these substitutions, we obtain

$$R_s = \frac{1}{4} n^{1/2} \left(1 - \frac{1 + k'_A}{1 + k'_B}\right) \tag{35}$$

Finally, since $\alpha = k'_B/k'_A$,

$$R_s = \frac{1}{4} n^{1/2} \left(\frac{\alpha - 1}{\alpha}\right) \left(\frac{k'_B}{1 + k'_B}\right) \tag{36}$$

the derivation of which is usually attributed to Purnell (63).

The three terms of equation 36 provide a good basis for discussing the factors that affect resolution (64, 65). However, there is another important variable that does not appear in the equation. It is the time of an analysis. In most practical analytical situations time is very important, and the resolution of two solutes must be accomplished in a reasonably short time. It seems proper, then, to discuss resolution within this context of "normalized time" (66–68).

The first term contains α, the selectivity or separation ratio. The equation indicates that a large value for α, and hence for $(\alpha - 1)/\alpha$, is desirable. In effect this means that the stationary phase should be chosen to show the greatest selectivity between the solutes A and B. This depends on the nature of the forces between the solute and the stationary phase, as already discussed.

Second, the partition ratio k' should be as large as possible so that the fraction $k'/(k' + 1)$ will be as large as possible. However, this statement must be modified on two accounts: little increase in $k'/(k' + 1)$ is achieved beyond a k' of about 5, and, the larger the value of k', the longer the time of analysis. In order to estimate the optimum k' in a minimum time, equations 36 and 37 can be combined to give a time expression:

$$t_R = n(k' + 1)\frac{H}{\bar{u}} \tag{37}$$

$$t_R = 16R_s^2\left(\frac{\alpha}{\alpha - 1}\right)^2 \frac{(k' + 1)^3}{(k')^2}\frac{H}{\bar{u}} \tag{38}$$

Then for minimum time, t_R, the term $(k' + 1)^3/(k')^2$ should be at a minimum.

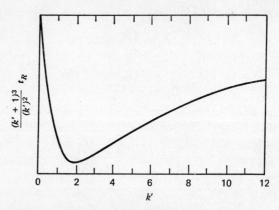

Figure 16. Effect of k' on retention time.

A plot of this term versus k' is shown in Figure 16 and goes through a minimum at $k' = 2$. This conclusion was reached by Grushka, but Guiochon found that a $k' = 3$ was optimal. In a recent joint paper (68) they conclude that the real optimum is between 2 and 3. However, they also conclude that the analysis time varies by only about 10% over the range of 1.5 to 5 in k' values. Clearly, for most work this range is reasonable. (Note, however, that k' is also a factor in the determination of n which will be discussed later.) An examination of GC literature reveals, however, that most analyses are run at much higher k' values.

It is instructive to compare the relative importance of α and k' in determining resolution. The values of α can vary from just over 1 up to, say, 2 (at which point the pair would probably be well separated). Hence $(\alpha - 1)/\alpha$ can vary from, for example, 0.001 up to 1, or a range of 10^3. By comparison, if k' varies from 1 to 50, $k'/(k' + 1)$ varies from 0.5 to nearly 1.0, or a range of only 0.5. Clearly, then, α is the more important variable for improving resolution.

Returning to equation 36, the third component is n, the number of theoretical plates, which is equal to L/H. According to the equation, R_s increases as n increases, but only by the 1/2 power. And, if n is increased by lengthening the column, the time of analysis increases. In this case, it will take a column four times as long to double the resolution; Guiochon (69) has found empirically that analysis time is approximately proportional to the 3/2 power of the column length. Thus an increase in column length by four times will increase the analysis time to eight times in order to double the resolution. By and large, longer columns are not the preferred method of improving resolution.

The other method of increasing the column efficiency, n, is to prepare and pack a better column. The van Deemter equation tells us how to get a maximum n or a minimum H:

$$H = 2\lambda d_p + 2\psi \frac{D_G}{\bar{u}} + \frac{8}{\pi^2} \frac{k'}{(k' + 1)^2} \frac{d_f^2}{D_L} \bar{u} = A + \frac{B}{\bar{u}} + C\bar{u} \qquad (39)$$

Typical plots of H versus $\bar{u}$ are shown in Figure 17. Clearly, the choice of optimum conditions is quite complex (70, 71) and only the most obvious parameters can be considered here. (The discussion assumes that n is a constant for a given column, but in fact it varies somewhat for each compound, as equation 39 would indicate.)

From Figure 17 it can be seen that there is an optimum flow or linear velocity. However, it must be remembered that the flow varies throughout the column depending on the P_i/P_0 ratio, so that the actual range of flows could be, for example, between X and Y in Figure 17b.

In practice a gas chromatograph is usually operated at a velocity greater than the optimum in order to decrease analysis time with relatively little loss of efficiency since the curve is often fairly flat on the high-velocity side. Scott and Hazeldean (72) have introduced the term "optimum practical gas velocity" (OPGV), shown as point Z in Figure 17b where the HETP tends to become linear.

Because of the emphasis on high-speed work, the B term of the rate equation does not contribute much to H since it becomes less important as $\bar{u}$ is increased. Nevertheless, equation 39 shows that the way to keep B minimal is to choose a gas phase like nitrogen, which gives small diffusion coefficients D_G. A lower H value can be obtained with nitrogen than with helium, but at a decreased velocity, as shown in Figure 17c. However, the H-curve for nitrogen rises more steeply with increased velocity than it does for helium, so helium is usually preferred for high-speed work. (The choice of the carrier gas is often influenced by the detector being used.) The tortuosity factor ψ ranges from 0.4 to 0.9 and is discussed more fully by Giddings (73).

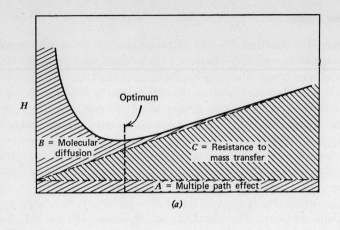

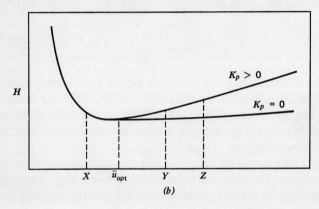

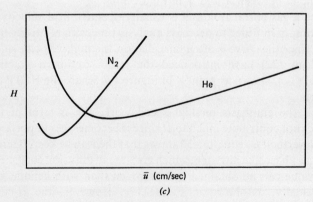

Figure 17. Typical van Deemter plots.

The A-term is flow independent and is minimized if the particle size d_p is small as well as the packing characteristic λ. Some typical values of λ were given in Chapter 8, which showed that it was smaller when d_p is large; that is, large particles can be packed more uniformly than small ones. However, the product of λ and d_p gives smaller values when d_p is small. But if d_p is too small, the pressure drop across the column becomes so large that a practical upper limit is reached. Obviously λ is also minimized if the range of particle sizes is narrow. Overall, a good size for particles is 100 to 120 mesh.

The C-term, or mass-transfer term, is the most important one in high-speed work. To keep it minimal, the film thickness d_f should be small, and the diffusion coefficient in the stationary liquid, D_L, should be large. Very few diffusion coefficients are available, but as a general rule high values can be expected in low-viscosity liquids, which are thereby preferred. The film thickness is more complex since it depends on the nature of the particles and the film. Low values mean low loaded columns and high β values. But the lower limit depends on the minimum amount necessary for complete coverage and the activity of the (inert) solid support. It has been shown that about 0.3% liquid phase is necessary to cover the active sites on Chromosorb P (74). Further details are given by Giddings (75).

We have already shown that a k' value of between 1.5 and 5 is optimal according to equation 38. In the third term the dependence on k' is $k'/(k' + 1)^2$, and the values decrease as k' increases. However, an increase in k' represents a proportionate increase in analysis time, and a plot of k' versus $k'/(k' + 1)^2$ shows that little is gained beyond a k' value of about 15. Obviously the choice of the best k' value for optimum H and R_s is quite complex and cannot be considered apart from the other variables such as K_p, β, and time.

In summary, for optimum resolution in minimum time, the best column should be prepared, consistent with equation 39, k' should be between 1.5 and 5, and the most selective liquid phase should be sought in order to get the best α value.

All of the factors that go into the preparation of a good GC column have been considered by Scott from an experimental point of view but in light of the theories just discussed (16). For those applications that warrant care and detailed consideration, his approach is probably the best. Swingle and Rogers (77) have shown how an on-line computer can be used to reduce the time necessary to follow Scott's procedure.

QUANTITATIVE ANALYSIS

Of all the separation methods used in the analytical laboratory, GC is most readily adapted for quantitative analysis. For that reason, a brief survey of quantitative methods for GC is included here. It is probable that as high-pressure LC develops, it will find greater use for quantitative analysis and the basic principles set forth here will apply equally well to LC.

Quantitative methods can be divided into three steps:

1. Generation of the analog signal
2. Conversion of analog signal to digital data
3. Relating of digital data to sample composition.

The first step was discussed for the TCD and FID in the section on detectors and will not be considered further. The other two are common procedures and hence will be discussed only briefly. The December 1967 issue of the *Journal of Gas Chromatography* contains 11 papers, which represent a good cross section of the subject (*78*). See also the review by Novak (*79*).

Analog-to-Digital Conversion

Normally chromatograms are analog representations of the type that has been presented, and the data must be converted to a digital form for calculation. The common methods are the following:

1. Peak height
2. Electromechanical (disk) integration
3. Triangulation
4. Planimeter
5. Electronic integrator
6. Cut-and-weigh
7. Computer
8. Tape system

Despite its widespread use, peak-height measurement is not highly recommended even though it is simple. The other methods are standard techniques for integrating the areas under peaks. They vary in precision, time required for the integration, and cost of the equipment. Each has to account for sloping base lines, nonsymmetrical peak shapes, and nonresolved peaks. For the manual methods (Nos. 3, 4, and 6) the recommendations of the Institute of Petroleum (*5*) are useful as well as the general papers by Condal–Bosch(*80*) and Ball, Harris, and Habgood (*81*). A more complete discussion of both manual and automated methods has been written by Johnson (*82*). In addition, the papers from the first (1972) International Symposium on Computer Techniques have been published in *Chromatographia* (*83*).

Sample Quantitation

Gas chromatography is a technique in which the experimental conditions cannot be reproduced exactly from day to day or run to run. Consequently accurate work requires the use of standards; the internal standard method is generally preferred. Furthermore, GC detectors do not respond in the same way for all compounds, so response factors should be used. Unfortunately, response factors for TCD and FID cannot be predicted with sufficient accuracy, so empirical response data must be obtained. The most complete summaries have been published by Dietz (84) and Rosie and Barry (85). Care should be exercised in using the values to note if they are for weight percent or mole percent and if they are to be multiplied times the area or divided into it.

The four standard methods for quantitation are the following:

1. Internal standard
2. Standard addition
3. Internal normalization
4. Direct calibration

The only special requirement in adapting them for GC analysis is that the response factors be used. Further details can be obtained from quantitative and instrumental analysis textbooks or from Rosie and Barry (85). A list of suggested internal standards for GC is also available (86).

MISCELLANEOUS TOPICS

Literature

From the beginning, the literature on GC has been compiled, reviewed, and abstracted, which makes retrieval of information easier than in most fields. Two abstracting services have been available since the 1950s. One is provided by volunteers and members of the GC Discussion Group of the Institute of Petroleum (87) and is available to members quarterly in paper and annually in hardback. Publication is relatively slow, but the cost is low. The other is a commercial venture (88) that began as notched cards but has been converted to a conventional monthly publication.

Two bibliographies of early GC work have been published. Over 15,000 references are cited in the two volumes by Signeur (89), which cover the literature to 1965. The other bibliography covers the period from 1952 to 1962

(90). It was updated on an annual basis by publications in the December issue of the *Journal of Gas Chromatography* for the years 1963 to 1966 (through November only). The total number of abstracts is nearly 13,000.

Beginning in 1960 *Analytical Chemistry* has included a separate biennial review on GC, of which the latest appeared in April 1974 (91). Finally, there are two important compilations of GC data. Reference has already been made to the one by McReynolds (42); the other one is published by ASTM (92).

Special Topics

Flow Programming. Temperature programming, already discussed, is a very useful and widely used technique. Similar in principle, but less widely used, is flow programming, which has been adequately described in two review articles (93). One difference between flow and temperature programming is that retention times have an inverse *linear* relationship to flow and an inverse *logarithmic* relationship to temperature.

Preparative. GC. Analytical chromatographs can accept only small samples, but they have been modified for larger columns and samples, and special instruments have been built to handle large preparative-scale separations (94). This subject also includes collection efficiency, which is important in preparative work.

Derivatization. Samples that are insufficiently volatile to be chromatographed directly can be reacted chemically to form volatile derivatives. Some of the derivatives and applications are listed in Table 16.

Table 16. Some Common Volatile Derivatives for GC

Derivative	Reagent	Application
Methyl esters	Diazomethane, BF_3-methanol	Acids
Silyl ethers	N,O-Bis(trimethylsilyl)acet-amide, trimethylsylilimi-dazole	Sugars, steroids, sterols, alcohols, acids, amides, amines (95)
Fluoroacetylacetonates	Trifluoroacetylacetone, hexafluoroacetylacetone	Metals (96)
Fluoroacyl derivatives	Trifluoroacetic anhydride	alcohols, amines, amino acids

EVALUATION

Gas chromatography has found wide application, from fixed gases to non-volatile solids that are derivatized or pyrolyzed. It is fast, efficient, and relatively inexpensive. The advantages and disadvantages are summarized in Table 17.

Table 17. Evaluation of GC

Advantages	Disadvantages
1. Efficient, selective, and widely applicable 2. Fast 3. Inexpensive and simple 4. Easily quantitated 5. Requires only a small sample 6. Nondestructive	1. Samples have to be volatile 2. Not suitable for thermally labile samples 3. Fairly difficult for large samples (preparative) 4. Theory inadequate, so some trial and error may be required

REFERENCES

1. A. T. James and A. J. P. Martin, *Biochem. J.* **50**, 679 (1952).

2. S. Kenworthy, J. Miller, and D. E. Martire, *J. Chem. Educ.* **40**, 541 (1963).

3. IUPAC, *Pure Appl. Chem.* **11**, 177 (1960); *ibid.*, **8**, 553 (1964).

4. ASTM, *Gas Chromatographic Terms and Relationships*, E355-68, 1968.

5. *J. Inst. Pet.* **53**, 367 (1967).

6. *Chromatographia* **1**, 153 (1968).

7. IUPAC, Information Bulletin No. 15, *Recommendations for Nomenclature in Chromatography*, February 1972.

8. H. M. McNair and C. D. Chandler, *J. Chromatogr. Sci.* **11**, 454 (1973).

9. M. Novotny, L. Blomberg, and K. D. Bartle, *J. Chromatogr. Sci.* **8**, 390 (1970) and references cited therein.

10. L. S. Ettre, J. E. Purcell, and K. Billeb, *J. Chromatogr.* **24**, 335 (1966) and references cited therein.

11. J. G. Nikelly, *Anal. Chem.* **45**, 2280 (1973).

12. L. S. Ettre, *Am. Lab.*, 2 [12], 28–39 (December 1970).

13. G. Guiochon in *Advances in Chromatography*, Vol. 8, J. C. Giddings and R. A. Keller (eds.), Dekker, New York, 1969, p. 179.

14. L. S. Ettre, *Open Tubular Columns*, Plenum Press, New York, 1965.

15. D. M. Ottenstein, *J. Gas Chromatogr.* **1**[4], 11 (April 1963).

16. D. M. Ottenstein, *J. Chromatogr. Sci.* **11**, 136 (1973).

17. D. E. Durbin, *Anal. Chem.* **45**, 818 (1973).

18. R. A. Keller, *J. Chromatogr. Sci.* **11**, 188 (1973); J. R. Mann and S. T. Preston, Jr., *ibid.*, 216; R. S. Henly, *ibid.*, 221.

19. A. G. Altenau, R. E. Kramer, D. J. McAdoo, and C. Merritt, Jr., *J. Gas Chromatogr.* **4**, 96 (1966); W. Fiddler and R. Doerr, *J. Chromatogr.* **21**, 481 (1966).

20. D. C. Locke, *J. Chromatogr. Sci.* **11**, 120 (1973); also see B. L. Karger and E. Sibley, *Anal. Chem.* **45**, 740 (1973).

21. R. J. Leibrand, *J. Gas Chromatogr.* **5**, 518 (1967); P. G. Jeffry and P. J. Kipping, *Gas Analysis by Gas Chromatography*, 2nd ed., Pergamon Press, New York, 1972.

22. O. L. Hollis, *Anal. Chem.* **38**, 309 (1966).

23. S. F. Spencer, *Anal. Chem.* **35**, 592 (1963).

24. D. David, *Gas Chromatographic Detectors*, Wiley–Interscience, New York, 1974.

25. A. E. Lawson and J. M. Miller, *J. Gas Chromatogr.* **4**, 273 (1966).

26. A. B. Richmond, *J. Chromatogr. Sci.* **9**, 92 (1971).

27. T. Johns and A. C. Stapp, *J. Chromatogr. Sci.* **11**, 234 (1973).

28. P. Bocek and J. Janak, *Chromatogr. Rev.* **15**, 111 (1971).

29. A. T. Blades, *J. Chromatogr. Sci.* **11**, 251 (1973).

30. M. Krejci and M. Dressler, *Chromatogr. Rev.* **13**, 1 (1970).

31. C. H. Hartmann, *Anal. Chem.* **43**[2], 113A (1971).

32. J. T. Walsh and D. M. Rosie, *J. Gas Chromatogr.* **5**, 232 (1967).

33. J. E. Lovelock, *Anal. Chem.* **35**, 474 (1963); W. A. Aue and S. Kapila, *J. Chromatogr. Sci.* **11**, 255 (1973).

34. A. Karmen, *J. Chromatogr. Sci.* **7**, 541 (1969).

35. M. C. Bowman and M. Beroza, *Anal. Chem.* **40**, 1448 (1968).

36. D. M. Coulson, in *Advances in Chromatography*, Vol. 3, J. C. Giddings and R. A. Keller (eds.), Dekker, New York, 1966, p. 197; *Am. Lab.* **1**[5] 22 (May 1969).

37. E. S. Kovats, *Helv. Chim. Acta* **41**, 1915 (1958); and in *Advances in Chromatography*, Vol. 1, J. C. Giddings and R. A. Keller (eds.), Dekker, New York, 1965, p. 229.

38. L. S. Ettre, *Anal. Chem.* **36** [8], 31A (July 1969).

39. G. Schomburg and G. Dielmann, *J. Chromatogr. Sci.* **11**, 151 (1973).

40. G. D. Mitra and N. C. Saha, *J. Chromatogr. Sci.* **8**, 95 (1970).

41. R. A. Hively and R. E. Hinton, *J. Gas Chromatogr.* **6**, 203 (1968); L. S. Ettre and K. Billeb, *J. Chromatogr.* **30**, 1 (1967); N. C. Saha and G. D. Mitra, *J. Chromatogr. Sci.* **8**, 84 (1970); P. G. Robinson and A. L. Odell, *J. Chromatogr.* **57**, 11 (1971).

42. W. O. McReynolds, *Gas Chromatographic Retention Data*, Preston Technical Abstracts Co., Evanston, Ill., 1966.

43. L. S. Ettre and W. H. McFadden (eds.), *Ancillary Techniques of Gas Chromatography*, Wiley–Interscience, New York, 1969.

44. J. O. Walker and C. J. Wolf, *J. Chromatogr. Sci.* **8**, 513 (1970); S. G. Perry, in *Advances in Chromatography*, Vol. 7, J. C. Giddings and R. A. Keller (eds.), Dekker, New York, 1968, p. 221.

45. J. T. Walsh and C. Merritt Jr., *Anal. Chem.* **32**, 1378 (1960).

46. S. F. Sarner, G. D. Pruder, and E. J. Levy, *Am. Lab.* **3**[10], 57 (1971).

47. D. G. Paul and G. R. Umbreit, *Research and Development*, p. 100 (May 1970); C. E. Bennett, L. W. DiCave, Jr., D. G. Paul, J. A. Wegener, and L. J. Levase, *Am. Lab.* 3[5], 67 (1971).

48. D. I. Rees, *Talanta* 16, 903 (1969).

49. B. L. Karger, *Anal. Chem.* 39[8], 29A (1967).

50. S. H. Langer and R. J. Sheehan, in *Advances in Analytical Chemistry and Instrumentation*, Vol. 6, J. H. Purnell (ed.), Interscience, New York, 1968, p. 289.

51. L. Rohrschneider, *Z. Anal. Chem.* 170, 256 (1959); and in *Advances in Chromatography*, Vol. 4, J. C. Giddings and R. A. Keller (eds.), Dekker, N.Y., 1967, p. 333.

52. L. Rohrschneider, *J. Chromatogr.* 22, 6 (1966).

53. W. R. Supina and L. P. Rose, *J. Chromatogr. Sci.* 8, 214 (1970).

54. W. O. McReynolds, *J. Chromatogr. Sci.* 8, 685 (1970).

55. J. H. Purnell, in *Gas Chromatography—1966*, A. B. Littlewood (ed.), Institute of Petroleum, London, 1967, p. 3.

56. H. Kelker and E. von Schivizhoffen, in *Advances in Chromatography*, Vol. 6, J. C. Giddings and R. A. Keller (eds.), Dekker, New York, 1968, p. 247.

57. E. Gil-Av and D. Nurok, in *Advances in Chromatography*, Vol. 10, J. C. Giddings and R. A. Keller (eds.), Dekker, New York, 1974, p. 99.

58. W. E. Harris and H. W. Habgood, *Talanta* 11, 115 (1964).

59. J. C. Giddings, *J. Chem. Educ.* 39, 569 (1962).

60. W. E. Harris and H. W. Habgood, *Programmed Temperature Gas Chromatography*, Wiley, New York, 1966.

61. L. Mikkelsen, in *Advances in Chromatography*, Vol. 2, J. C. Giddings and R. A. Keller (eds.), Dekker, New York, 1966, p. 337.

62. W. E. Harris and H. W. Habgood, *J. Gas Chromatogr.* 4, 144, 168, 217 (1966).

63. J. H. Purnell, *J. Chem. Soc.*, 1960, 1268.

64. B. L. Karger, *J. Chem. Educ.* 43, 47 (1966).

65. L. S. Ettre, *Am. Lab.* 2[12], 28 (1970).

66. B. L. Karger and W. D. Cooke, *Anal. Chem.* 36, 985, 991 (1964).

67. S. Hawkes, *J. Chromatogr. Sci.* 7, 526 (1969).

68. E. Grushka and G. Guiochon, *J. Chromatogr. Sci.* 10, 649 (1972) (see also the references therein).

69. G. Guiochon, *Anal. Chem.* 38, 1020 (1965).

70. J. C. Giddings, in *Advances in Analytical Chemistry and Instrumentation*, Vol. 3, C. N. Reilly (ed.), Wiley–Interscience, New York, 1964, p. 315.

71. I. Halasz and E. Heine in *Advances in Analytical Chemistry and Instrumentation*, Vol. 6, J. H. Purnell (ed.), Wiley–Interscience, New York, 1968, p. 153.

72. R. P. W. Scott and G. S. F. Hazeldean, in *Gas Chromatography—1960*, R. P. W. Scott (ed.), Butterworths, Washington, 1960, p. 144.

73. J. C. Giddings, *Dynamics of Chromatography*, Dekker, New York, 1965, p. 243.

74. P. Urone, Y. Takahashi, and G. H. Kennedy, *Anal. Chem.* 40, 1130 (1968).

75. J. C. Giddings, *Dynamics of Chromatography*, Dekker, New York, 1965, p. 136.

76. R. P. W. Scott, in *Advances in Chromatography*, Vol. 9, J. C. Giddings and R. A. Keller (eds.), Dekker, New York, 1970, p. 193.

77. R. S. Swingle and L. B. Rogers, *Anal. Chem.* **44**, 1415 (1972).

78. *J. Gas Chromatogr.* **5**, 595–646 (1967).

79. J. Novak, in *Advances in Chromatography*, Vol. 11, J. C. Giddings and R. A. Keller (eds.), Dekker, New York, in press.

80. L. Condal-Bosch, *J. Chem. Educ.* **41**[4], A235 (1969).

81. D. L. Ball, W. E. Harris, and H. W. Habgood, in *Separation Techniques in Chemistry and Biochemistry*, R. A. Keller (ed.), Dekker, New York, 1967, p. 153.

82. H. W. Johnson, Jr., in *Advances in Chromatography*, Vol. 5, J. C. Giddings and R. A. Keller (eds.), Dekker, New York, 1968, p. 175.

83. *Chromatographia* **5**, 61–211 (1972).

84. W. A. Dietz, *J. Gas Chromatogr.* **5**, 68 (1967).

85. D. M. Rosie and E. F. Barry, *J. Chromatogr. Sci.* **11**, 237 (1973).

86. W. W. Fike, *J. Chromatogr. Sci.* **11**, 25 (1973).

87. C. E. H. Knapman, ed., *Gas Chromatography Abstracts*, Applied Science Publishing, Barking, Essex, England, 1972. Earlier annual edition published by Elsevier and Butterworths.

88. *Gas Chromatography Literature*, Vol. 6, Preston Technical Abstracts Co., Niles, Ill., 1972.

89. A. V. Signeur, *Guide to Gas Chromatography Literature*, Plenum, New York, 1964 and 1967.

90. S. T. Preston, Jr., and G. Hyder, *A Comprehensive Bibliography and Index to the Literature on Gas Chromatography*, Preston Technical Abstracts Co., Evanston, Ill., 1965.

91. S. P. Cram and R. S. Juvet, Jr., *Anal. Chem.* **46**, 101R (1974).

92. O. E. Schupp, III, and J. S. Lewis, *Gas Chromatographic Data Compilation*, ASTM–AMD 25A and Supplement 1, 2nd ed., AMD 25A S 1, American Society for Testing and Materials, Philadelphia, 1967 and 1971.

93. R. P. W. Scott, in *Advances in Analytical Chemistry and Instrumentation*, Vol. 6, Wiley, New York, 1968, p. 271; L. S. Ettre, L. Mazor, and J. Takacs, in *Advances in Chromatography*, Vol. 8, J. C. Giddings and R. A. Keller (eds.), Dekker, New York, 1969, p. 271.

94. G. W. A. Rijnders, in *Advances in Chromatography*, Vol. 3, J. C. Giddings and R. A. Keller (eds.), Dekker, New York, 1966, p. 215; D. T. Sawyer and G. L. Hargrove, in *Advances in Analytical Chemistry and Instrumentation*, Vol. 6, C. N. Reilly (ed.), Wiley, New York, 1968, p. 325.

95. A. E. Pierce, *Silylation of Organic Compounds*, Pierce Chemical Co., Rockford, Ill., 1968; *Handbook of Silylation*, Pierce Chemical Co., Rockford, Ill., 1972; G. Brittain, *Am. Lab.* **1**[5], 57 (1969).

96. R. W. Moshier and R. E. Sievers, *Gas Chromatography of Metal Chelates*, Pergamon Press, Oxford, 1965.

SELECTED BIBLIOGRAPHY

Burchfield, H. P., and E. E. Storrs, *Biochemical Applications of Gas Chromatography*, Academic Press, New York, 1962. Much broader coverage than title suggests.

Dal Nogare, S., and R. S. Juvet, Jr., *Gas–Liquid Chromatography*, Interscience, New York, 1962.

Ettre, L. S., *Practical Gas Chromatography*, Perkin–Elmer Co., Norwalk, Conn., 1973.

Ettre, L. S., and A. Zlatkis (eds.), *The Practice of Gas Chromatography*, Interscience, New York, 1967.

Jones, R. A., *Gas–Liquid Chromatography*, Academic Press, New York, 1970.

Littlewood, A. B., *Gas Chromatography*, 2nd ed., Academic Press, New York, 1970.

McNair, H. M., and E. J. Bonelli, *Basic Gas Chromatography*, 5th ed., Varian Aerograph, Walnut Creek, Calif., 1969.

Purnell, H., *New Developments in Gas Chromatography*, Vol. II, Wiley-Interscience, New York, 1973.

Schupp, O. E., III, in *Technique of Organic Chemistry*, Vol. XIII, E. S. Perry and A. Weissberger (eds.), Interscience, New York, 1968.

Tranchant, J. (ed.), *Practical Manual of Gas Chromatography*, Elsevier, Amsterdam, 1969.

LIQUID CHROMATOGRAPHY IN COLUMNS: GENERAL

13

Although liquid chromatography in columns is the original form of chromatography introduced by Tswett at the turn of the century, the technique has come into prominence only recently. The main reason for the revived interest is the development of theory and instrumentation that has made possible faster and more efficient analyses. This development followed, and was catalyzed by, the development of GC (*1*). As one would expect, there are many similarities between LC in columns and GC. Since GC is covered in Chapter 12, LC will be approached from a perspective that facilitates a comparison of the two techniques. In theory, they are very similar; in practice, they are complementary.

Liquid chromatography in columns can be subdivided into separate techniques based on the different mechanisms by which separations are effected. These specific techniques are covered in Chapter 14; the general principles and instrumentation are covered here.

Chapter 11 contains a classification of chromatographic techniques, including the distinction in bed configuration—column or plane surface (Table 1, B and C). The planar methods (TLC and PC) are covered in Chapter 15, which includes comparisons of the two types of bed. Another distinction between LC methods is the pretreatment of the bed; it can be dry, or it can be wetted with mobile liquid. This can also be a determining factor in the method used to pack the columns. Usually the planar methods are carried out on dry beds and the column methods on wetted beds.

The first LC separations were performed in columns; the mobile liquid moved through the bed under the force of gravity. This technique is called *gravity flow* or *traditional* or *classical* LC; a brief description follows. The rest of this chapter and the next one deal with modern, or high-pressure, LC.

CLASSICAL LC

Classical column LC was virtually the only type used prior to the mid-1940s, so chromatography texts published before about 1950 treat only this type.

Books on modern LC did not appear until 1971; those published between 1950 and 1971 cover classical LC, PC, and TLC, and possibly some GC. Thus all of the classical books on chromatography predate modern LC techniques.*

The procedure is a simple one. A tube or column, usually made of glass, with a diameter of 1 to 5 cm and up to 50 cm long, is packed with a solid stationary phase such as silica gel. A frit or plug of glass wool at the bottom retains the packing, which may be used dry or may be wetted with the mobile liquid. If it is wet, the liquid level is allowed to drop to the top of the bed (but never below it), and the sample is added. Small amounts of solvent are added to wash the sample onto the column, keeping it in a small volume. Additional solvent is added and flows through the column under the force of gravity until the sample components are separated in the column or are eluted from it. In the latter case, small fractions are taken for analysis by such techniques as refractive index or absorption spectroscopy.

The elution procedure is slow and may take several hours; the identification (and/or quantitation) of solutes in large numbers of fractions is tedious. Usually the column is not reused. The only advantages of the classical procedure are its simplicity and low cost, but these do not usually justify its use. Student experiments from the *Journal of Chemical Education* provide further information and examples (2).

HIGH-PRESSURE LC APPARATUS

A general description of chromatographic apparatus is given in Chapter 11, and a good summary of modern LC equipment is available in Kirkland's book (3). A comparison of the commercially available equipment has been published (4). A summary is presented here of the major parts of a liquid chromatograph for general use, although instruments for a specific technique may differ somewhat.

Pumping Systems

The following items (in order of their location in the liquid stream) are included in this section: reservoir and degass system, gradient device, pump, pulse damper, and precolumn (LLC). The reservoir should be made of an inert material, such as glass or stainless steel. Its capacity is determined by the analysis; 1 liter should be sufficient. Solvents in which oxygen is soluble, such as water, may need to be degassed before use. This can be done before the

---

* See, for example, the Selected Bibliography in Chapter 11.

sample is put into the reservoir, or the degassing capability can be built into the reservoir. The common methods for degassing are (1) distilling, (2) heating slightly with stirring, and (3) evacuating. Problems with dissolved gases usually arise from small bubbles in the detector.

Gradient Devices. If it is desirable to change the composition of the mobile liquid during a run, a device is necessary for producing the gradient. An exponential gradient can be generated simply at atmospheric pressure by allowing a solvent (usually the more polar one) to flow by gravity into a stirred mixing vessel that contains the initial solvent and feeds the pump.

More flexibility is possible if the solvents are pumped individually into a mixing chamber of small volume (a length of tubing may suffice). The desired gradient can be obtained by programming the pumps, and it can vary from linear to exponential in either solvent. Needless to say, these systems are fairly expensive.

Pumps. Suitable pumps should be capable of delivering up to 10 ml of solvent per minute at pressures of at least 1000 psig. Four types are listed in Table 1, including suppliers and some characteristics of their pumps. This list is intended to be representative of the pumps in use, and not a complete listing of all available suppliers.

The pressurized reservoir is the simplest type and produces an uninterrupted pulseless flow. However, this type of pump does have a number of disadvantages, including limited pressure, limited volume, unsuitability for

Table 1. Pumps for High-Pressure LC

Type	Examples	Maximum Pressure (psig)	V or P^a	Volume Limit (ml)
Pressurized reservoir	Varian Aerograph	1000	P	500
Reciprocating	Milton Roy	1000/3000	V	None
	Orlita	4875	V	None
	Whitey	1000	V	None
	Waters	6000	V	None
Motor-driven syringe	Perkin–Elmer	3000	V	250
	Varian Aerograph	5000	V	250
	ISCO	2000	V	375
Pneumatically driven syringe	Haskell Engineering	3300	P	70/stroke

^{a} V = constant volume, P = constant pressure.

gradient work, and the tendency for the pressurizing gas to dissolve in the solvent. The Varian pump is compartmentalized to minimize gas diffusion into the solvent volume, and other manufacturers have taken similar precautions to keep the interface small. Other possibilities include the use of a mercury barrier and a collapsible plastic bottle (5). Finally, this type of pump produces a constant pressure, which is less desirable than a constant flow. If a change in pressure occurs in the system downstream from the pump, the flow will change and will probably go undetected. This situation is very undesirable.

The reciprocating pump is very popular, but it produces a pulsed flow (except for the Waters unit) and requires some type of damping. Its advantages are that it can be used with unlimited volume of solvent, it delivers constant volume, and its low internal volume makes it ideal for gradient work. The Waters pump is somewhat different from the others in that two pistons are operated out of phase by two elliptical gears so that the composite flow is pulseless.

Both the motor-driven and the pneumatically driven syringe pumps are pulseless. To avoid the limitation on volume, they can be used in pairs or they can be designed with a very fast refill stroke. They are not desirable for gradient work.

Pulse Dampers. A wide variety of dampers have been used, but most require a large solvent volume and/or a solvent volume that is static and out of the flowing stream. Such an arrangement is undesirable for gradient work and causes problems when the mobile-phase solvent is changed. Typical examples are pressure gauges of the Bourdon tube type, piston dampers, hydrodynamic accumulators, and lengths of flexible coiled tubing. Huber (6) has used an in-line Bourdon tube and a capillary restriction downstream; Freeman and Zielinski (7) designed a combination of accumulators and restrictions in analogy with an electrical RC filter. One of the best and least expensive dampers is a 10-ft length of 0.030-in. (inside diameter) Teflon tubing that has been compressed flat in a vise and wound on a mandrel (8). The total volume is about 0.25 ml, and it is completely in-line.

Precolumn. In LLC the solvent must be presaturated with the stationary liquid phase, so it is not stripped off the analytical column during use. Even if the solvent in the reservoir is presaturated, a precolumn is highly desirable. The dimensions are not important, except that too large a volume is undesirable; the packing in the precolumn contains the same stationary phase as the analytical column and is held on a fairly large-particle-size support, so it does not create any significant backpressure. In some cases the volume in the precolumn can aid in damping the pulses from a reciprocating pump.

Sampling

Sample introduction in LC is similar to that in GC. Either a sample loop or syringe injection through a septum can be used. The latter is simpler and most common, but it is more difficult in LC due to the high pressure of the inlet, which causes septa to rupture more frequently and requires the use of higher pressure syringes to prevent blowback. Furthermore, if on-column injection is attempted, the dead volume must be kept very small, and needles often become plugged with column packing. A superior injection port design uses two septa (*9*).

An alternative is to stop the flow, allow the pressure to subside, inject the sample, and then restore the pressure and flow. This is called *stop-flow injection* and is suitable for use in LC because of the low diffusion coefficients in liquids. It does have the disadvantage that column packings that are compressible may shift in position when the pressure is changed.

A sampling problem can arise from low solubility of the sample in the mobile liquid. Normally a sample would be introduced as a solution in the mobile liquid. If the solubility is low, samples as large as several hundred microliters may be necessary. Should this size provide an insufficient concentration, another solvent must be used. It should be readily soluble in the mobile liquid to facilitate rapid mixing, and its elution peak must not interfere with the sample peaks. In LLC one must also be careful that the sample solvent does not strip off the stationary phase from the head of the column packing.

Columns (*10*)

Column materials, geometry, and connections are much more critical in LC than in GC. Significantly higher efficiencies are obtained with Trubore glass (heavy wall for high pressure) and precision-bore stainless steel (*11*). Apparently very smooth inner walls are required for best efficiency.

Straight columns are preferred for reasons to be discussed later. Connections between columns and between columns, and the injection port and detector must be of very low volume. Connections that involve a change in inside diameter should be specially constructed to maximize mixing. Consequently LC columns are usually short (1 m or less), and several are connected in series if greater length is required. A more recent suggestion is that columns be coiled in a "figure 8" (*12*).

The most commonly used column inside diameters are 2 to 3 mm. However, some workers have used columns of about 11 mm inside diameter and obtained better efficiencies in LSC (*13*). The present theory is that in columns this wide (and about 50 cm long) a solute injected in the center of the column packing does not reach the column wall before it is eluted due to the small lateral diffusion rate. Thus there is no wall effect if only the center of the eluting zone is detected. Such columns have been called *infinite diameter columns.*

Temperature control is sometimes necessary, especially in LLC, and elevated temperatures are often preferable in ion-exchange separation. Water jackets are used for small columns; larger columns are more conveniently accommodated in larger forced air ovens.

Stationary Phases and Supports. A wide variety of materials is used as solid stationary phases and as solid supports for liquid stationary phases, and the number is still increasing rapidly. Majors has compiled a comprehensive summary as of 1972 (*14*). Most of the discussion of stationary supports is presented in the next chapter; only a general introduction is included here.

In a packed LC column there should be no deep pools of stagnant mobile phase and no long diffusion paths (*15*). Consequently only the very small particle sizes of conventional porous packings are suitable. Common particle-size ranges are 37 to 50 μm, 25 to 37 μm, and less than 40 μm. Even smaller sizes are becoming popular as packing techniques improve; some packings are available that are nominally 10 and even 5 μm in diameter.

Another type of packing has been developed to achieve the same desirable results. These are packings with a solid core and a porous layer that is one-thirtieth to one-fortieth of the solid core diameter. They are also referred to as superficially porous or controlled surface porosity (CSP) packings. The thin layer of stationary phase limits the capacity of the column, so only small samples can be used.

Since developments in this field continue, a complete tabulation is not possible. However several summaries present the state of the art as of 1972, including commercial products and trademarks (*14, 15*). Reference *14* also includes a few scanning electron micrographs, as does reference *16*. One of the earliest commercial CSP products was described by Kirkland (*17*) and is now known as Zipax.

Another type of special packing material has organic groups chemically bonded to the surface (*18*). These were first reported in 1968 for modified porous silica beads (*19*); since it was thought that a monolayer is formed, with the organic group protruding from the silica surface, they were called *brushes.* Chemical groups have also been bonded to the porous layer-type materials. Some of these are of the brush type, and others are crosslinked polymers that form a skin on the solid core. The latter materials are referred to as *pellicular* supports. The first ones were ion-exchange materials, but newer ones include hydrocarbon polymer and polyamide coatings as well as silica, alumina, and molecular sieves (20). Hence the term "pellicular" is also applied to materials of the porous layer type. Since much of the information about the commercial material is proprietary, a strict classification is difficult.

Locke (18) lists the following types of bonded support and trade names:

1. Brush or ester packings (Durapak)
2. Silicone polymer (Bondapak, Vydak, Permaphase)
3. Grignard or organolithium bonded phases
4. Pellicular (Pellionex, Zipax-SAX)

Table 2 represents a different classification scheme and provides surface areas and pore sizes for some. Not included in this listing are the regular ion-exchange resins and the polymers used in gel-permeation chromatography, which are covered in the next chapter.

The method of packing the support into a column varies widely and depends on the type of support. Some materials, such as ion-exchange resins and gel-permeation packings, must be packed wet as a slurry since they swell when solvated. The classical procedure is to pour the slurry into a column, allow the particles to settle, and continue until the column is full. A slurry method for ion-exchange packings that produces higher efficiency columns has been described (21). For other materials that can be packed either dry or wet, most chromatographers prefer dry packing. The column can be agitated in a variety of ways (see, for example, reference 22) similar to the procedures used in GC. Sie and von den Hoed (23) have studied the various dry and wet techniques, and show photographs of columns exhibiting undesirable size segregation; Snyder's review of column efficiencies (24) also contains a good summary of packing procedures. Kirkland (22) showed that a "tap-fill" method is best for Zipax–type supports.

The most efficient columns have a plate height of about 0.1 mm. Two methods for packing microparticle porous silica have been reported to produce columns of that efficiency. One is a dry packing method in which the packing is admitted in small quantities and then tamped (25), and the other is a wet "balanced-density" slurry method under 5000-psi pressure, (26). Obviously such columns are very tightly packed, and thus more pressure is required to achieve a given flow. A total evaluation of column performance should include not only efficiency (H) but also permeability (pressure required) and speed.

Mobile Phase.   The general requirements of the mobile phase are that it have the proper "polarity" for the desired separation, low viscosity, high purity and stability, and compatibility with the detection system. It must also wet the stationary phase and dissolve the sample. If the sample is to be recovered, a

Table 2. Some Packing Materials for LC

Type	Trade Name	Manufacturer[a]	Surface Area (m²/g)	Pore Size (Å)
Porous particles:				
Silica	Porasil, Spherosil (six types)	P & W	2–6 to 350–500	100 to 1500
	Li Chrosorb	E	>200	60
	Microsil	R	400	50
Diatomite	—	—	10	—
Alumina	—	—	>200	—
Bonded porous particles, silica				
	Durapak	W	50 to 100	200–400
Surface porosity (pellicular) particles, silica core				
	Corasil	W	7 and 14	—
	Pellosil	R	4 and 8	—
	Vydac	S	12	57
	Zipax	D	1	1000
Bonded surface porosity particles, silica core				
	Durapak/ Corasil	W	7	—
	Bondapak/ Corasil	W	7	—
	Permaphase	D	<1	—
	Vydac reverse phase	S	12	—

[a] Manufacturers: P = Pechiney-St. Gobain; W = Waters Associates; E = EM Laboratories; R = Reeve Angel; D = Dupont; S = Separations Group.

volatile mobile liquid is often chosen so it can be easily evaporated. Toxicity and flammability are also considerations, depending on the amount of care taken.

Detectors

As in GC, the detector is usually an integral part of a modern LC instrument. About a dozen LC detectors have been described, but the most popular ones are the ultraviolet absorption type, followed by the refractive index detector. These are briefly described here, but good reviews on LC detectors can provide additional information (*27–32*).

Ultraviolet Absorption   The principle of operation of a UV detector for LC is the same as for conventional UV detectors. The sample cell must accept a flowing liquid stream and have a small volume (10 to 20 μl). The design is important (*30*). The photometer can be either single beam or double beam, but in the latter case a static reference cell can be used.

Until 1973 virtually all of the commercially available UV detectors were either single or at most dual wavelength. The source was a low-pressure mercury lamp, and the lines used were 254 and 280 nm. In some instruments a phosphor is used to convert the 254-nm radiation to 280 nm to achieve dual-wavelength operation. Typical sensitivities are 0.01 and 0.005 absorbance unit full scale (AUFS).

Newer instruments incorporate monochromators and can be used over a range of wavelengths. In most cases a standard UV spectrophotometer or monochromator has been adapted by designing a suitable cell and condensing the beam size, if necessary.

Use of a UV detector requires a UV-absorbing sample and a nonabsorbing mobile phase. Ideally they are very sensitive and can be used for gradient elution. Temperature regulation is not usually required, and they are less sensitive to flow changes (such as the pulses from a pump) than refractive index detectors. A general summary has been written by Bakalyar (*33*).

A detector similar in principle is the one commonly used to detect amino acids. A reagent, ninhydrin, is mixed with the column effluent to produce a color that can be monitored at 570 nm. Thus the detector is a visible spectrometer or colorimeter with modifications for the reagent mixing. Such a detector has not been widely used, but reactions that develop color or UV absorption are well known, so it is likely that it will become more popular.

Refractometers.   Two types of differential refractometer are available for LC. The more conventional design is a deflection type (*34*), and the other is a Fresnel type (*35*). Optical diagrams are shown in Figure 1.

In the deflection type a beam of light passes through the cell, which contains both the reference and sample streams, and its deflection is determined by the difference in refractive index (RI) between the two. As the beam changes location on the detector, a signal is generated.

The Fresnel type is based on Fresnel's law: the fraction of light reflected at a glass–liquid interface is proportional to the angle of incidence and the relative refractive indices of the substances. For LC the angle of incidence is adjusted so that it is slightly less than the critical angle, and the detector

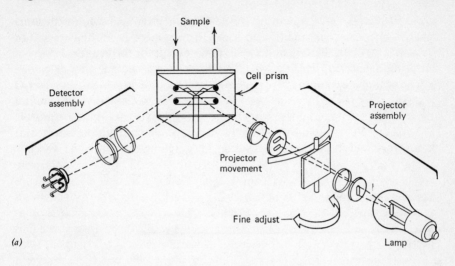

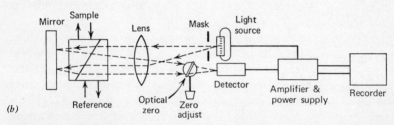

Figure 1. Optical diagrams of refractometric detectors for LC: (*a*) Fresnel type. By permission of Laboratory Data Control Co.; (*b*) deflection type. By permission of Waters Associates, Inc., Milford, Mass.

responds to the varying intensity of light striking it. As shown in Figure 1, this detector has two cavities (sample and reference) and two light-sensitive elements in a dual photodetector. This type of cell has a smaller volume than the deflection type.

Unfortunately, RI is very temperature sensitive, and even the use of a differential cell does not remove the requirement for good temperature control if high sensitivity is desired. For maximum sensitivity of 10^{-5} to 10^{-6} refractive index unit full scale (RIUFS), the temperature should be maintained at about $\pm 0.001°$C. Consequently this type of detector is seldom used at full sensitivity. A further limitation is its lack of applicability for gradient elution; only in rare circumstances can a series of solvents be found that will produce the desired gradient and have the same RI. The major advantage of the RI detector is its wide range of applicability.

Other Detectors. A list of other LC detectors is given in Table 3 with some specific references (in addition to those in references *27* through *32*). The principles of operation are quite well known except for the last three (heat of adsorption detector, Christiansen effect detector, and moving-wire detector).

The heat of adsorption (*40*) or microadsorption detector (MAD) consists of a small cavity packed with an adsorbent (often the same material packed in the column) and containing a temperature-sensing device, such as a thermistor. When a sample is eluted from the column and passes through the detector, it is adsorbed and then desorbed, resulting in exothermic and then endothermic

Table 3. Some Other LC Detectors

Detector	Reference	Commercially Available	Advantage of Use
Fluorimeter	—	Yes	Sensitive
Conductimeter	*36*	Yes	Aqueous solutions
Dielectric constant detector	*37*	No	—
Polarograph	*38*	No	—
Radioactivity detector	*39*	No	Specific
Heat of adsorption detector	*40*	Yes	—
Christiansen effect detector	*41*	Yes	Similar to RI
Moving-wire detector	*42*	Yes	Universal

changes. The thermistor senses the temperature changes and produces a second-differential type of curve. Good temperature control on the detector is required.

The Christiansen effect detector (CED) was introduced in 1973 (*41*). It is based on the principle that a solid immersed in a liquid of the same RI is not visible. Actually, if the dispersions of the two substances differ, colored light is produced under collimated illumination, and this effect was used by Christiansen to make color filters. The LC detector consists of a cell cavity packed with an insoluble solid having the same RI (at one wavelength) as the mobile liquid. Colored light is transmitted when the mobile liquid passes through the cell, but it changes when a sample appears in the eluent. The properties of the CED are quite similar to those of the RI detector.

Early in the 1960s some chromatographers attempted to adapt the FID used in GC for use in LC. The basic idea was to catch some of the column effluent on a moving wire, pass it through a low-temperature oven to flash off the solvent, and then through a high-temperature pyrolyzing oven. The pyrolysis products from each solute are swept into an FID and detected as in GC. This detector has become known by a variety of names, including "moving-wire detector," "FID for LC," "phase-transformation detector," and "solute-transport detector." It is a universal detector as long as the solutes are relatively nonvolatile, and it can be used with gradient elution.

A modification of this system suggested in 1970 (*42*), has resulted in several new commercial instruments (see, for example, reference *43*). Instead of being pyrolyzed, the solutes are first oxidized to carbon dioxide and then catalytically reduced to methane before being swept to the FID. In this case the signal depends on the carbon content of the sample, and it is more sensitive for some oxygenated compounds. Sensitivity can be further improved if the efficiency of effluent collection can be improved (*44*). Its disadvantages are its large size and the problems associated with the mechanical system.

Finally, it should be mentioned that fraction collectors are commercially available that permit a type of semiautomatic form of detection. They can be set to collect effluent into small containers automatically, but they need to be analyzed manually later.

THEORY

A few theoretical subjects applicable to most LC processes should be discussed. It is often useful to approach them from a perspective that compares LC with GC.

Diffusion in the Mobile Phase

In the comparison between GC and LC in Chapter 11 it was noted that diffusion coefficients for gases are about 10^5 times as large as for liquids; some typical values were given in Chapter 4. Also, the viscosity of liquids is about 100 times as large for liquids as for gases. Consequently the nature of the packed bed and the speed of analysis in LC are different from those in GC. The slow diffusion in the mobile liquid can cause significant broadening if long diffusion paths are available, so LC packings have either a small diameter or a porous layer on a solid core. The high viscosity of liquids requires the use of high inlet pressures in order to get reasonably fast flow rates and fast analyses, but the use of small packing particles produces the opposite effect. Hence in LC one must consider the interrelationships among type of packing, particle size, method of packing, sphericity of the particles, permeability of the packing, viscosity of the liquid, and speed of analysis. For further discussion we need to consider a definition of permeability.

Permeability.   Permeability is defined as

$$\kappa = \frac{\eta L \epsilon_T \bar{u}}{\Delta P} \tag{1}$$

where η is the viscosity, L is the column length, ϵ_T is the total porosity, $\bar{u}$ is the average mobile-phase velocity, and ΔP is the pressure drop across the column. For a given type of packed column the pressure must be increased to increase the velocity; to maintain a constant velocity, it must be increased if the viscosity is increased, the length is increased, or the permeability is decreased.

The Kozeny–Carmen equation provides a relationship between permeability and particle diameter d_p:

$$\kappa = \frac{d_p^2}{180} \left[\frac{\epsilon_I^3}{(1 - \epsilon_I)^2} \right] \tag{2}$$

where ϵ_I is the interstitial porosity, which is typically about 0.42. In that case, equation 2 reduces to

$$\kappa = \frac{d_p^2}{1000} \tag{3}$$

Since permeability varies with the square of the particle diameter, much is to be gained from using small particles.

An equation giving the effect of κ on retention time can be obtained by substituting equations 1 and 2 into an equation from Table 2 in Chapter 11:*

$$t_R = n(1 + k') \frac{H}{\bar{u}} \tag{4}$$

We get

$$t_R = \frac{\eta \epsilon_T (1 + k') L^2}{\kappa \, \Delta P} = \frac{\eta \epsilon_T (1 + k')}{\Delta P} \left(\frac{L}{d_p} \right)^2 \tag{5}$$

If all other factors are constant, the retention time is inversely proportional to the permeability. In LC work these equations should be used to confirm that a newly packed column has the degree of permeability it should and that there are no obstructions in the system. Further, they are used to compare the various packings and packing methods to arrive at the best compromise among the operating variables.

Coiling of Columns.   In GC, columns are usually coiled to permit small ovens to be used since this has no adverse effect on their performance. In LC, however, the low diffusivities in the liquid mobile phase retard lateral mass

---

* In Chapter 11 this equation was given in terms of V_R^0 rather than t_R, but the principle is the same; only the units must be changed.

transfer, and coiling can have an adverse effect. It is believed that in a coiled column the molecules, on the inside bend move faster than those on the outside, and the latter also have a longer path to travel. Consequently a solute concentration profile is set up across the column, with the molecules on the inside (of the bend) ahead of those on the outside.

It has been found that coiled columns can be used in LC if the direction of the bend is reversed, giving a "figure 8" configuration. A good discussion with recommendations has been published (*12*). On the other hand, it has also been reported that gel-permeation columns can be coiled with no loss in performance (*45*).

The Rate Equation in LC

The general form of the chromatographic rate equation was given in Chapter 8, and its use in GC was considered in Chapter 12. The main difference in LC is that the terms that involve mobile-phase diffusion and mass transfer are more important than they are in GC for the reasons just discussed. Hence the simple three-term equation of van Deemter is inadequate for LC. Peak dispersion in the mobile phase has been extensively investigated (*46, 47*) but is beyond the scope of this text.

The shape of the H versus $\bar{u}$ plot is shown in Figure 2. It differs from the parabolic shape of the GC curve. The broken-line portion of the curve is seldom shown in experimental plots because it occurs at such low velocities. Some complete curves are shown in work by Huber and co-workers (*48, 49*). These papers should be consulted for further information about the complete rate equation of Huber.

Giddings (*50*) has presented evidence of a coupling effect in the mobile phase, and his rate equation is equally complex; additional details were given in Chapter 8. However, for most practical purposes, a simple equation proposed by Snyder is a close approximation to the experimental data between 0.1 and 10 cm/sec (*51*):

$$H = D(u)^{0.4} \tag{6}$$

where D is a constant for a given column that Snyder has shown empirically to be related to the particle diameter d_p:

$$D = 18(d_p)^{0.8} \tag{7}$$

In summary, one can see from Figure 2 that a decrease in velocity always produces an increase in performance under normal conditions, which is not true in GC. Like in GC, however, the velocity chosen is often much higher in order to decrease the analysis time.

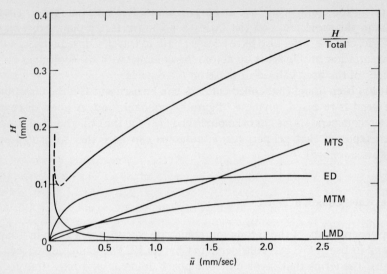

Figure 2. Typical plot of the rate equation for liquid chromatography (MTS = mass transfer in the stationary phase; ED = eddy diffusion; MTM = mass transfer in the mobile phase; LDM = longitudinal molecular diffusion). Reprinted from J. F. K. Huber, *J. Chromatogr. Sci.* **7**, 85, (1969) by courtesy of the Preston Technical Abstracts Co.

Optimization

Chapter 12 contains an extensive discussion about optimization in GC based on the equation

$$R_s = \frac{1}{4} \frac{(\alpha - 1)}{\alpha} \frac{k'}{k' + 1} n^{1/2} \tag{8}$$

This same equation applies to LC optimization, and to a certain extent this discussion is a repeat of the earlier one, except that the emphasis is on LC and its special problems.

First, we must review other relevant equations. Since our optimization includes minimization of the time required, equation 4 can be combined with equation 8 to give

$$t_R = 16 R_s^2 \left(\frac{\alpha}{\alpha - 1}\right)^2 \frac{(k' + 1)^3}{(k')^2} \frac{H}{\bar{u}} \tag{9}$$

For much of the LC literature another equation is used; it is derived from equation 4 and the definition of "effective theoretical plates," n_{eff} (Chapter 11):

$$n_{\text{eff}} = n \left(\frac{k'}{k' + 1}\right)^2 \tag{10}$$

Some authors refer to $(k'/(k' + 1))$ as Q and to equation 10 as $n_{eff} = nQ^2$. Note that this Q is (in our symbols) the quantity $(1 - R_R)$.] The combined equation is

$$\frac{n_{eff}}{t_R} = \frac{(k')^2}{(1 + k')^3} \frac{\bar{u}}{H} \tag{11}$$

Obviously the number of effective plates per second, n_{eff}/t_R, is a useful measure of optimized performance and time. For best results it should be maximized.

We have already seen that the term $(k')^2/(1 + k')^3$ goes through a maximum at a k' value of 2 and that for most work k' can vary from 1.5 to 5 for good results. Snyder has published a highly recommended two-part paper giving a rapid method for selecting the best experimental conditions for LC (52). Another recent paper that includes some practical aspects of time normalization is by Grushka (53).

Karger suggests that the product of the permeability κ and the number of effective plates per second, n_{eff}/t_R, is one indication of the efficiency of a column and the pressure needed to attain it (54). Table 4 is from his work and shows a comparison of some column types. The best columns using this criterion are those with the highest values in the last column. The GC columns are clearly superior, although the porous layer–type LC columns approach packed GC columns in "plates per second."

At this point it is appropriate to recall the theoretical comparison between GC and LC presented in Chapter 11. The present discussion is much more

Table 4. Comparison of Column Types in GC and LC[a]

Column Type	$\dfrac{n_{eff}}{t_R}$	$\left(\dfrac{n_{eff}}{t_R}\right) \kappa \times 10^7$ (cm^2/sec)
Gas chromatography:		
Classical packed, d_p 130 μm	10	240
Open tubular, d_c 0.25 mm	25	4700
Packed capillary, d_p 10 μm	40	270
Porous layer beads, d_p 90 μm	50	75
Liquid chromatography:		
Classical, d_p 150 μm	0.02	0.045
Silica, d_p 20 μm	2	0.08
Porous layer beads, d_p 27 μm	8	0.68

[a] From Karger (54).

realistic in presenting the state of the art. Liquid chromatography is not practiced at the very high pressures or the very long times necessary to achieve the large number of plates predicted theoretically.

MISCELLANEOUS TOPICS

The fact that the mobile phase competes with the stationary phase for the sample adds another dimension to achieving selectivity, which is not possible in GC with its inert carrier gas. This effect can be exploited still further in some cases by programming the mobile liquid during a run, a technique known as *gradient elution*. It is used in GSC and ion-exchange separations, and to a limited extent with bonded phases in LLC. Essentially, gradient elution is to LC what programmed temperature is to GC. One significant difference is that in programmed temperature the temperature change occurs simultaneously over the entire column, whereas in gradient elution the new solvent (or solvent mixture) is introduced only at the column inlet. For this reason some chromatographers would prefer that specific terms were associated with specific techniques, but this is not the case. In fact, the process of moving a heater from the column inlet to the outlet is known as chromatothermography in GC, and it is this technique that most closely resembles gradient elution in operation.

Temperature effects are difficult to predict in LC, and programmed temperature is seldom used. For most work it is sufficient to maintain a constant temperature near ambient. Elevated temperatures are usually used (1) if the separation process is kinetically controlled and is slow at ambient temperature or (2) to reduce the solvent viscosity.

In conclusion, there are two problems in LC that do not occur in GC and should be mentioned. First is the difficulty in determining V_M, the column dead volume. There is no simple marker for LC like air, which is used in GC. Some alternatives are suggested by Halasz (55). Second is the fact that the base widths of adjacent peaks may not be equal as they are in GC. This can cause an error in the calculation of resolution since the equation we used assumes equality.

Recycle

One method of increasing the efficiency of a separation that has not been mentioned is to recycle the sample through the same column. This is an attractive alternative to the use of long columns since lower pressures are required. However, in each cycle the sample does pass through the pump, so it is necessary to use a small-volume reciprocating pump. The process is most commonly used in gel-permeation work and has been shown to be effective (56).

Field Flow Fractionation (57)

In 1966 a technique was proposed that resembles LC. It is called *field flow fractionation*, and it achieves separations by differential migration in a one-phase system. A lateral field that can be thermal, magnetic, centrifugal, and so on, is coupled with an axial liquid flow. It is mentioned here to illustrate the kind of processes that can be devised from basic principles and from a comparison of different separation techniques.

EVALUATION

The advantages and disadvantages of LC are similar to those for GC. The major difference is that LC can be used with nonvolatile samples. A complete summary is given in Table 5.

Table 5. Evaluation of LC

Advantages	Disadvantages
1. Applicable to nonvolatile samples	1. Slower than GC
2. Efficient, selective, and widely applicable	2. Experimentally more complex than GC
3. Can be quantitated	
4. Requires only a small sample	
5. Nondestructive	

REFERENCES

1. C. Karr, Jr., E. E. Childers, and W. C. Warner, *Anal. Chem.* **35,** 1291 (1963); J. C. Giddings, *ibid.* **35,** 2215 (1963).
2. J. M. Bohen, M. M. Joullie, F. A. Kaplan, and B. Loew, *J. Chem. Educ.* **50,** 367 (1973); I. B. Ruppel, Jr., F. L. Cuneo, and J. G. Krause, *ibid.* **48,** 635 (1971); S. J. Romano, *ibid.* **47,** 478 (1970); H. H. Strain and J. Sherma, *ibid.* **46,** 476 (1969); S. Marmor, *ibid.* **42,** 272 (1965).
3. R. A. Henry, in *Modern Practice of Liquid Chromatography*, J. J. Kirkland (ed.), Wiley-Interscience, New York, 1971.
4. C. D. Chandler and H. M. McNair, *J. Chromatogr. Sci.* **11,** 468 (1973).

5. S. G. Perry, R. Amos, and P. I. Brewer, *Practical Liquid Chromatography*, Plenum, New York, 1972, p. 167.

6. J. F. K. Huber, *J. Chromatogr. Sci.* **7**, 85 (1969).

7. D. H. Freeman and W. L. Zielinski, Jr., *NBS Technical Note 589*, National Bureau of Standards, Washington, D.C., 1971.

8. R. P. W. Scott and P. Kucera, *J. Chromatogr. Sci.* **11**, 83 (1973).

9. B. Pearce and W. L. Thomas, *Anal. Chem.* **44**, 1107 (1972).

10. J. J. Kirkland, *Anal. Chem.* **43**[12], 36A (1971).

11. J. J. Kirkland, *J. Chromatogr. Sci.* **7**, 361 (1969).

12. H. Barth, E. Dallmeier, and B. L. Karger, *Anal. Chem.* **44**, 1726 (1972).

13. J. J. DeStefano and H. C. Beachell, *J. Chromatogr. Sci.* **10**, 654 (1972), and references cited therein.

14. R. E. Majors, *Am. Lab.* **4**[5], 27 (May 1972).

15. R. E. Leitch and J. J. DeStefano, *J. Chromatogr. Sci.* **11**, 105 (1973).

16. M. DeMets and A. Lagasse, *J. Chromatogr. Sci.* **8**, 272 (1970).

17. J. J. Kirkland, *J. Chromatogr. Sci.* **7**, 7 (1969); *Anal. Chem.* **41**, 218 (1969).

18. D. C. Locke, *J. Chromatogr. Sci.* **11**, 120 (1973).

19. I. Halasz and I. Sebastian, *Angew Chem. Int. Ed.* **8**, 453 (1969).

20. H. M. McNair and C. D. Chandler, *Anal. Chem.* **45**, 1117 (1973).

21. C. D. Scott, *J. Chromatogr.* **42**, 263 (1969).

22. J. J. Kirkland, *J. Chromatogr. Sci.* **10**, 129 (1972).

23. S. T. Sie and N. von den Hoed, *J. Chromatogr. Sci.* **7**, 257 (1969).

24. L. R. Snyder, *J. Chromatogr. Sci.* **7**, 352 (1969).

25. J. F. K. Huber, F. F. M. Kolder, and J. M. Miller, *Anal. Chem.* **44**, 105 (1972).

26. R. E. Majors, *Anal. Chem.* **44**, 1722 (1972); W. Strubert, *Chromatographia* **6**, 50 (1973).

27. C. D. Conlon, *Anal. Chem.* **41**, 107A (1969).

28. J. F. K. Huber, *J. Chromatogr. Sci.* **7**, 172 (1969).

29. H. Veening, *J. Chem. Educ.* **47**, A549, A675 (1970).

30. S. H. Byrne, Jr., in *Modern Practice of Liquid Chromatography*, J. J. Kirkland (ed.), Wiley-Interscience, New York, 1971, p. 95.

31. J. Polesuk and D. G. Howery, *J. Chromatogr. Sci.* **11**, 226 (1973).

32. H. Veening, *J. Chem. Educ.* **50**, A429, A481, A529 (1973).

33. S. R. Bakalyar, *Am. Lab.* **3**[6], 29 (1971).

34. J. L. Waters, *Am. Lab.* **3**[5], 61 (1971).

35. E. S. Watson, *Am. Lab.* **1**[9], 8 (September 1969).

36. A. Ford and C. E. Meloan, *J. Chem. Educ.* **50**, 85 (1973).

37. R. Vespalec and K. Hana, *J. Chromatogr.* **65**, 53 (1972); A. Jackson, *J. Chem. Educ.* **42**, 447 (1965); H. Poppe and J. Kuysten, *J. Chromatogr. Sci.* **10**[4], 16A (1972).

38. J. G. Koen, J. F. K. Huber, H. Poppe, and G. D. Boef, *J. Chromatogr. Sci.* **8**, 192 (1970).

39. A. M. van Urk-Schoen and J. F. K. Huber, *Anal. Chim. Acta* **52**, 519 (1970).

40. M. N. Monk and D. N. Raval, *J. Chromatogr. Sci.* **7**, 48 (1969).

41. Personal communication, Gow-Mac Instrument Co., Madison, N.J., 1973.

42. R. P. W. Scott and J. G. Lawrence, *J. Chromatogr. Sci.* **8**, 65 (1970).

43. M. H. Pattison, *Am. Lab.* **4**[5], 55 (1972).

44. J. H. van Dijk, *J. Chromatogr. Sci.* **10**, 31 (1972); V. Pretorius and J. F. J. van Rensburg, *J. Chromatogr. Sci.* **11**, 355 (1973).

45. L. R. Whitlock, R. S. Porter, and J. F. Johnson, *J. Chromatogr. Sci.* **10**, 437 (1972).

46. D. S. Horne, J. H. Knox, and L. McLaren, in *Separation Techniques in Chemistry and Biochemistry*, R. A. Keller (ed.), Dekker, New York, 1967, p. 97.

47. J. H. Knox and J. F. Parcher, *Anal. Chem.* **41**, 1599 (1969).

48. J. F. K. Huber and J. A. R. J. Hulsman, *Anal. Chim. Acta* **38**, 305 (1967).

49. J. F. K. Huber, *J. Chromatogr. Sci.* **7**, 85 (1969).

50. J. C. Giddings, *Anal. Chem.* **35**, 1338 (1963).

51. L. R. Snyder, *J. Chromatogr. Sci.* **7**, 352 (1969).

52. L. R. Snyder, *J. Chromatogr. Sci.* **10**, 200, 369 (1972).

53. E. Grushka, *J. Chromatogr. Sci.* **10**, 616 (1972).

54. B. L. Karger, in *Modern Practice of Liquid Chromatography*, J. J. Kirkland (ed.), Wiley–Interscience, New York, 1971, p. 47.

55. I. Halasz, in *Modern Practice of Liquid Chromatography*, J. J. Kirkland (ed.), Wiley–Interscience, New York, 1971, p. 328.

56. K. J. Bombaugh, W. A. Dark, and R. F. Levangie, *J. Chromatogr. Sci.* **7**, 42 (1969).

57. J. C. Giddings, *Sep. Sci.* **1**, 123 (1966); G. H. Thompson, M. N. Myers, and J. C. Giddings, *ibid.* **2**, 797 (1967); J. C. Giddings, *J. Chem. Educ.* **50**, 667 (1973).

SELECTED BIBLIOGRAPHY

Baumann, F., and N. Hadden, *Basic Liquid Chromatography*, Varian Aerograph, Walnut Creek, Calif., 1971.

Brown, P. R., *High Pressure Liquid Chromatography, Biochemical and Biomedical Applications*, Academic Press, New York, 1972.

Karger, B. L., L. R. Snyder, and C. Horvath, *An Introduction to Separation Science*, Wiley–Interscience, New York, 1973.

Kirkland, J. J. (ed.), *Modern Practice of Liquid Chromatography*, Wiley–Interscience, New York, 1971.

Perry, S. G., R. Amos, and P. I. Brewer, *Practical Liquid Chromatography*, Plenum, New York, 1972.

Snyder, L. R. and J. J. Kirkland, *Introduction to Modern Liquid Chromatography*, Wiley–Interscience, New York, 1974.

LIQUID CHROMATOGRAPHY IN COLUMNS: INDIVIDUAL TECHNIQUES

14

Chromatographic processes can be classified according to the states of the two phases or according to the mechanisms by which the separations are effected. The techniques described in this chapter represent a mixture of these two classification systems, but they are the names by which the techniques are most commonly known. They are liquid–solid (LSC), liquid–liquid (LLC), ion exchange (IEC), and gel permeation (GPC) or exclusion chromatography. Each section describes the theory, stationary phase, mobile phase, and other topics unique to that technique.

LIQUID–SOLID CHROMATOGRAPHY

Liquid–solid chromatography is the technique used originally by Tswett and the one most commonly associated with "classical" LC over the last 70 years. Several monographs on LSC have been written (1, 2), including a very comprehensive one by Snyder (3). The two chapters he has written in Kirkland's book (4) are also worthy of special mention. Any of the classical LC books (see Chapter 11) also contained considerable information about LSC, although much of it is not relevant to modern techniques.

This discussion is limited to LSC in columns, but the planar techniques (TLC and PC) are for the most part liquid–solid processes also, so the literature is somewhat mixed. Comparisons of the column and planar techniques are presented in Chapter 15.

Theory

Because the stationary phase is a solid, solute partitioning occurs at the liquid–solid interface by the mechanism defined as adsorption (see Chapter 5 for an introduction to adsorption). As we have seen, adsorption processes

230

often have nonlinear adsorption isotherms that can limit the separating efficiency in LSC. The phenomenon is less severe than it is in GSC since the liquid mobile phase is more effective than an inert carrier gas in occupying the most active sites or deactivating the surface. However, most mobile liquids are relatively nonpolar and do not provide adequate deactivation, so other means are often used.

The surfaces of chromatographic adsorbents are generally heterogeneous and there are differences in activity; thus the concept of covering up the most active sites with a deactivating agent is realistic. In many cases water is used. Snyder has found that best results (maximum linear capacity) are obtained with a water monolayer covering 50 to 100% of the solid surface. This requires up to 0.04 g of water per 100 m² of surface or about 4 to 15% water added to the chromatographic adsorbent.

Adding water to the system is less simple than it sounds. For reproducibility, the solid must first be carefully dried and then a measured amount of water added. Complications can arise from the adsorption of water from air by the chromatographic support. In fact, there is probably some adsorbed water on the surface of any active support, so some water will be present in all systems unless extreme care is taken to exclude it. Once water is in the system, it will come to equilibrium with the mobile liquid over a period of time. Thus, even at constant temperature, the water content of the liquid and the solid can vary until equilibrium is reached. In the meantime the column may appear to give inconsistent retention times and changing peak shapes. It would appear that most chromatographers do not pay close attention to this phenomenon in the preparation of their LS systems. Further details are given by Snyder (ref. 4, p. 224).

For simplicity, our discussions of the solute–solvent–solid interactions in LSC will omit consideration of a deactivator or moderator like water. Several models that can lead to useful mathematical relationships are possible (1), but only a brief discussion will be given.

The heterogeneous solid surface will have some average surface activity, and some mobile liquid molecules will adsorb on it. When a solute molecule appears in the mobile liquid, it may have some attraction for mobile liquid molecules. It will also be strongly attracted to the solid surface; however, in order to adsorb on it, it is necessary for it to dislodge a solvent molecule from the site. To represent this (solute) sorption and (solvent) desorption, Snyder has proposed the following relation:

$$S^0 - A_s \epsilon^0 \tag{1}$$

where S^0 is the adsorption energy of the solute, ϵ^0 is the adsorption energy (per unit area of standard activity) of the solvent, and A_s is the area of solid required by the adsorbed solute. The parameter ϵ^0 is also referred to as

"Snyder's eluent strength function"; A_s can be calculated by summing tabulated areas for the common molecular groups of which the solute is composed. Equation 1 can be related to the partition coefficient by the following relation

$$\log K_p = \log V_a + E_a(S^0 - A_s\epsilon^0) \tag{2}$$

where V_a is the "adsorbent surface volume" or the volume of an adsorbed monolayer of mobile liquid and E_a is the average surface activity of the solid.* This equation can be used to interpret the adsorption process in LSC and may in the future be used to predict conditions for separations.

For example, in comparing the effects of two different mobile liquids on a given solute and a given solid, it can be shown that the relative partition coefficients are

$$\log \frac{(K_p)_1}{(K_p)_2} = E_a A_s(\epsilon_2^0 - \epsilon_1^0) \tag{3}$$

The ratio of retention volumes in the two solvents would depend on their difference in eluent strength, since E_a and A_s would be constant. This concept is the basis for the selection of the mobile phase in LSC.

In addition, if we consider the separation of two solutes in each of these two solvents, the magnitude of their relative sizes, A_s, can be important. Table 1 gives some experimental data for two solutes on a column of 3.7% water on alumina (ref. 3, p. 191). It shows that this effect can be large enough to reverse the order of separation for two solutes in the two different systems.

The two parameters in equation 2 in widespread use are E_a and ϵ^0, particularly the latter. Some values for E_a as a function of the percentage of

Table 1. An Example of the Effect of ϵ^0 and A_s on LSC Separations[a]

		Relative V_R	
		Pentane	Benzene
Sample	A_s	$\epsilon^0 = 0.00$	$\epsilon^0 = 0.32$
1,2,4,5-Dibenzpyrene	15	365	0.6
2,6-Dimethylpyridine	8	46	2.0

[a] From reference 3, p. 191, by courtesy of Marcel Dekker, Inc.

* Snyder uses the symbol α, rather than E_a, in his work.

Table 2. Effect of Water Deactivator on the Activity of Large-Pore Silica[a]

Percentage of Water	E_a
0	0.83
0.5	0.79
1.0	0.75
2.0	0.71
4.0	0.70
7.0	0.69
10.0	0.69

[a] From reference *3*, p. 136, by courtesy of Marcel Dekker, Inc.

water added as a deactivator are given in Table 2 (ref. *3*, p. 136). Clearly, as the percentage of water is increased, the activity of the silica decreases.

Eluent strength parameters ϵ^0 for different solvents also depend on the solid with which they are being used. The relative values vary little, and Snyder has given a table of ϵ^0 values on alumina. He also notes that ϵ^0 on silica is equal to 0.77 of the value on alumina; since silica is the more common adsorbent, the converted values for silica are given in Table 3. We shall return to a consideration of solvents later in the chapter.

Stationary Phase

Normally the stationary phase used in LSC is an active solid, and silica is by far the most popular one used. For modern LC two types of silica are available: porous microbeads with diameters from 5 to 20 μm and the so-called porous layer beads, which have a surface layer of silica on a solid core. These were discussed in Chapter 13. It should be noted that some of these materials have relatively inactive surfaces and are intended only as solid supports for LLC (e.g., Liqua–Chrom and Zipax). Also available commercially are alumina and a porous layer polyamide (nylon). The two most popular solids, silica and alumina, differ in that the first is acidic and the second is usually basic.

Classically, a wide variety of materials of varying activity have been used. Table 4 lists the most common ones in order of increasing activity, but it is presented mainly for historical purposes. Most of the attention in LSC is on the mobile liquid.

Table 3. Solvent Properties of Some Liquids

Solvent	Estimated ϵ^0 (Silica)	δ	η	RI
Fluoroalkanes	−0.19	∼5.5	—	1.25
n-Pentane	0.00	7.1	0.23	1.358
Isooctane	+0.01	7.0	0.50	1.404
Cyclohexane	+0.03	8.2	1.00	1.427
Cyclopentane	+0.04	8.1	0.47	1.406
1-Pentene	+0.06	—	0.18	1.371
Carbon disulfide	+0.11	10.0	0.37	1.626
Carbon tetrachloride	+0.14	8.6	0.97	1.466
Xylene	+0.20	8.9	0.6–0.8	∼1.5
Isopropyl ether	+0.22	∼7.3	0.37	1.368
Isopropyl chloride	+0.22	∼8.4	0.33	1.378
Toluene	+0.22	8.9	0.59	1.496
Chlorobenzene	+0.23	9.5	0.80	1.525
Benzene	+0.25	9.2	0.65	1.501
Ethyl ether	+0.29	7.4	0.23	1.353
Chloroform	+0.31	9.3	0.57	1.443
Methylene chloride	+0.32	9.7	0.44	1.424
Methyl isobutyl ketone	+0.33	—	0.58	1.394
Tetrahydrofuran (THF)	+0.35	9.1	—	1.408
Ethylene dichloride	+0.38	9.7	0.79	1.445
2-Butanone (MEK)	+0.39	9.3	0.44	1.381
Acetone	+0.43	9.9	0.32	1.359
Dioxane	+0.43	10.0	1.54	1.422
Ethyl acetate	+0.45	9.6	0.45	1.370
Methyl acetate	+0.46	9.2	0.37	1.362
Amyl alcohol	+0.47	9.8	4.1	1.410
Dimethyl sulfoxide (DMSO)	+0.48	12.8	2.24	1.479
Nitromethane	+0.49	12.6	0.67	1.394
Acetonitrile	+0.50	11.7	0.37	1.344
Pyridine	+0.55	10.7	0.94	1.510
Propanol, n and i	+0.63	11.5	2.3	1.38
Ethanol	+0.68	12.7	1.20	1.361
Methanol	+0.73	14.4	0.60	1.329
Ethylene glycol	+0.86	14.7	19.9	1.427
Acetic acid	Large	—	1.26	1.372
Water	Large	21	1.00	1.333

Table 4. Solids Used as Stationary Phases in LSC (In Order of Increasing Activity)

1. Sucrose	9. Calcium phosphate
2. Starch	10. Magnesium carbonate
3. Inulin	11. Magnesium oxide
4. Magnesium citrate	12. Silica gel
5. Talc	13. Magnesium silicate (Florisil)
6. Sodium carbonate	14. Alumina
7. Potassium carbonate	15. Charcoal, activated
8. Calcium carbonate	16. Fuller's earth (kaolin-type clay)

Mobile Phase

Many of the common liquids used as mobile phases are listed in Table 3. The basic requirements are (1) proper strength or "polarity" (ϵ^0 or δ); (2) low viscosity; (3) compatibility with detector (e.g., lack of ultraviolet absorption); (4) stability and compatibility with active solid; and (5) volatility to facilitate solute recovery. Most of our attention will be focused on property 1 since it is the main parameter that can be varied to achieve a desired separation. The stationary phase is usually silica, as already mentioned.

The selection of the proper polarity solvent begins with an educated guess using the "polarity scales" in Table 3 or the older eluotropic series such as that in Table 5. If the elution takes too long, a stronger one is chosen, and vice versa.

However, linearity can be improved (tailing reduced) if a modifier or de-activator is added to the main solvent, as already discussed. Obviously, the addition of a polar modifier to the solvent will increase the solvent polarity and deactivate the stationary phase. The selection of a solvent mixture is more

Table 5. Liquids Used as Mobile Phases in LSC—Eluotropic Series (In Order of Increasing Polarity)

1. Petroleum ether	9. Ethyl ether
2. Cyclohexane	10. Ethyl acetate
3. Carbon tetrachloride	11. Acetone
4. Trichloroethylene	12. n-Propyl alcohol
5. Toluene	13. Ethanol
6. Benzene	14. Methanol
7. Dichloroethylene	15. Water
8. Chloroform	

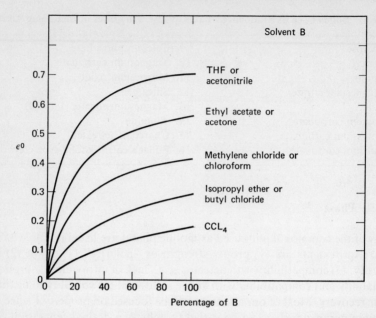

Figure 1.   Polarity (ϵ^0) of mixed solvents as a function of composition. Solvent A is heptane (Courtesy of the Perkin–Elmer Corp.)

difficult, but Snyder has provided ϵ^0 data for some mixtures (*3*); a brief discussion has also been provided by Yost and Conlon (*5*). Figure 1 is typical of the information available; it shows that different amounts of a variety of modifiers can be added to heptane to make mixtures of increased ϵ^0 value. According to the theory, any mixture with a given ϵ^0 value should produce the same chromatographic behavior for a given sample. [The term "isocratic" has been coined to describe a solvent (or solvent mixture) whose composition (and polarity) does not change during a run.]

As an example of this principle, Yost and Conlon (*5*) chromatographed acrylamide on a Sil-X column with four different mixtures of modified chloroform, each with an ϵ^0 of 0.53. The chromatograms are shown in Figure 2, and it can be seen that the retention time is nearly constant, as predicted. However, the use of ethanol (Fig. 2*b*) causes an undesirable peak broadening and would be unsuitable for use. This effect could not have been predicted and indicates the desirability of trying several solvent mixtures.

The use of an isocratic solvent will be successful only if the range of solute partition ratios (k') is between approximately 1 and 10. If the range is wider, a single solvent system will not provide the best separation in the least time, and gradient elution is recommended.

Gradient Elution. Gradient elution (changing the polarity, ionic strength, or pH during a run) has been shown to be preferable to coupling columns or to flow programming or temperature programming in LSC (*6*). Snyder and Saunders (*7*) have given a thorough discussion of the ways to optimize a solvent gradient. They conclude that, in an optimum solvent program, ϵ^0 should change by about 0.04 unit per column volume of solvent. Several recommended solvent programs are given in Table 6; they can be run step-wise or as a continually changing gradient. Obviously, series III would be easier to run as a continuous gradient, and the other two would be more suitable for stepwise operation. Very little difference has been reported for the two types of operation, and both are generally referred to as *gradient elution*.

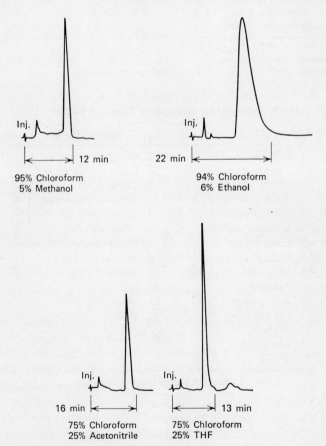

Figure 2. Comparison of four mobile-phase solvent mixtures each with an ϵ^0 value of 0.53. Conditions: sample, acrylamide; column, 50 cm × 3 mm Sil-X; flow, 1 ml/min. Courtesy of the Perkin–Elmer Corp.

Table 6. Three Eluotropic Series for Silica Columns[a–c]

ϵ^0	Eluotropic Series		
	I	II	III
0.00	Pentane	Pentane	Pentane
0.05	4.2% PrCl in pentane	3% CH_2Cl_2 in pentane	4% Benzene in pentane
0.10	10% PrCl in pentane	7% CH_2Cl_2 in pentane	11% Benzene in pentane
0.15	21% PrCl in pentane	14% CH_2Cl_2 in pentane	26% Benzene in pentane
0.20	4% Ether in pentane	26% CH_2Cl_2 in pentane	4% EtOAc in pentane
0.25	11% Ether in pentane	50% CH_2Cl_2 in pentane	11% EtOAc in pentane
0.30	23% Ether in pentane	82% CH_2Cl_2 in pentane	23% EtOAc in pentane
0.35	56% Ether in pentane	3% Acetonitrile in benzene	56% EtOAc in pentane
0.40	2% Methanol in ether	11% Acetonitrile in benzene	
0.45	4% Methanol in ether	31% Acetonitrile in benzene	
0.50	8% Methanol in ether	Acetonitrile	
0.55	20% Methanol in ether		
0.60	50% Methanol in ether		

[a] After Snyder (ref. *4*, p. 221).
[b] All percentages are by volume.
[c] Abbreviations: PrCl, isopropyl chloride; EtOAc, ethyl acetate.

Table 7. Solvents for Incremental Gradient Elution

Series I (*8*)	Series II (*9*)
1. *n*-Heptane	1. *n*-Heptane
2. Carbon tetrachloride	2. Carbon tetrachloride
3. Heptyl chloride	3. Chloroform
4. Trichloroethane	4. Ethylene dichloride ⎫
5. *n*-Butyl acetate	5. 2-Nitropropane ⎪
6. *n*-Propyl acetate	6. Nitromethane ⎬ Mixtures
7. Ethyl acetate	7. Propyl acetate ⎪
8. Methyl acetate	8. Methyl acetate ⎭
9. Ethyl methyl ketone	9. Acetone
10. Acetone	10. Ethanol
11. *n*-Propanol	11. Methanol
12. Isopropanol	12. Water
13. Ethanol	
14. Methanol	
15. Water	

Column-Reconditioning Solvents

1. Ethanol
2. Acetone
3. Ethyl acetate
4. Trichloroethane
5. Heptane

"Eluotropic series" is an older term used to describe a series of solvents that increases in polarity (ϵ^0).

Scott and Kucera (8, 9) have published two similar stepwise gradients that cover a wide range of polarity and can be used as general screening gradients (see Table 7). In series II solvents 3 through 7 are used in mixtures of two to five of these solvents. This series was based on k' values rather than ϵ^0 values. Each was carefully selected to keep the k' for a given solvent (chromatographed in the previous solvent) constant at about 0.32. Furthermore, in an effort to keep the changes in dispersion forces constant, the difference in molecular weight between consecutive solvents was held approximately constant at about 14, decreasing from carbon tetrachloride to water. A test chromatogram of 21 widely different solutes is shown in Figure 3.

One of the reasons for using so many solvents in a gradient is the displacement effect, which occurs when the mobile phase is drastically changed. We noted earlier that some solvent is always adsorbed on the solid surface. When a more polar solvent is introduced, it displaces the previous solvent from the surface and may cause the elution of a large enough quantity of solvent to produce a peak. The displacement effect also works on the adsorbed solutes, so some of them show very narrow peak widths resulting from displacement and not elution. This effect is minimized by using a large number of solvents and a gradual increase in solvent polarity.

Likely peak identity

1	Squalane	15	Quinine
2	Anthracene	16	Acetylsalicylic acid
3	Methyl stearate	17	Benzoic acid
4	Benzophenone	18	t–BOC leucine
5	Chloroaniline	19	t–BOC glycine
6	Nitroaniline	20	Alanine
7	p–Dinitrobenzene	21	Glucose
8	p–Nitrophenol		
9	Dihydrocholesterol		
10	Catechol		
11	Phenacetin		
12	Adenine		
13	Phenolphthalein		
14	EEDQ		

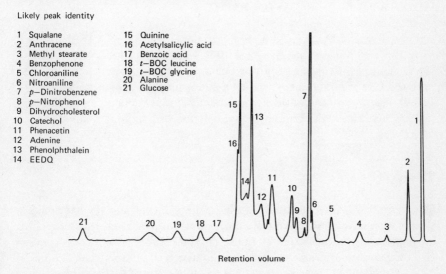

Retention volume

Figure 3. LSC Incremental gradient-elution chromatogram. Conditions: column, 50 cm × 5 mm inside diameter; packing, Bio–Sil A; flow rate, 0.5 ml/min; charge, 10 mg in 50 μl. Reprinted with permission from R. P. W. Scott and P. Kucera, *Anal. Chem.*, **45**, 749 (1973). Copyright by the American Chemical Society.

At the completion of a gradient elution run the column must be regenerated to restore it to its original polarity. The reverse sequence of solvents recommended by Scott and Kucera is included in Table 7. About 10 column volumes of each solvent are used.

Applications and Discussion

The LSC technique is best suited for the separation of compounds by type and by functional group. In some cases high selectivity is exhibited for isomer separations. The use of gradient elution facilitates the analysis of samples with widely varying polarities. A few examples are shown in Figures 4 to 6, which illustrate common types of adsorbents, solvents, and separation efficiencies.

The disadvantages of LSC include nonlinearity (tailing peaks) and the difficulty in reproducing stationary-phase activity, especially when deactivators are used. With isocratic LSC a solute may be irreversibly adsorbed due to the high activity of the stationary solid surface. Finally, although it adds versatility, gradient elution is not simple.

LIQUID–LIQUID CHROMATOGRAPHY

Compared with LSC, LLC is a new technique originating with the Nobel-prize work of Martin and Synge in 1941 (*12*). In principle it is much like liquid–liquid extraction, and in practice much of the column technology has followed GLC. Kirkland has written a good summary of modern LLC (*13*). In our discussion we consider permanently bonded phases as liquids and include them along with the conventional nonbonded stationary liquid phases.

Theory

The thermodynamics of liquid–liquid partition theory and regular solution theory were covered in Chapters 3 and 5, and will not be repeated. Detailed summaries have been written by Locke (*14*) and Huber (*15*). A longer review by Soczewinski (*16*) includes extensive discussion of intermolecular forces in LLC. A method has been suggested for the prediction of partition coefficients in LLC, and good correlation with static measurements has been achieved (*17*).

The rate theory of column efficiency was presented in Chapter 13. Emphasis was placed on the necessity for using small particles or porous layer materials.

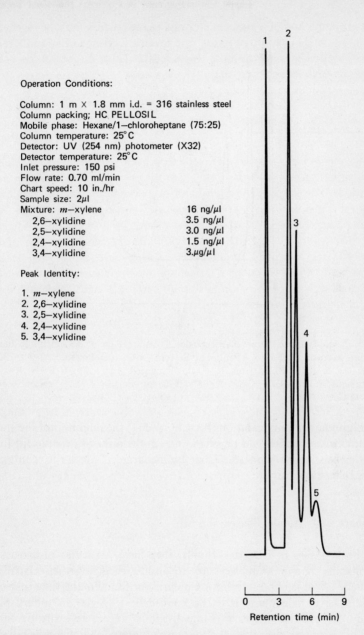

Operation Conditions:

Column: 1 m × 1.8 mm i.d. = 316 stainless steel
Column packing; HC PELLOSIL
Mobile phase: Hexane/1–chloroheptane (75:25)
Column temperature: 25°C
Detector: UV (254 nm) photometer (X32)
Detector temperature: 25°C
Inlet pressure: 150 psi
Flow rate: 0.70 ml/min
Chart speed: 10 in./hr
Sample size: 2µl

Mixture: m–xylene	16 ng/µl
2,6–xylidine	3.5 ng/µl
2,5–xylidine	3.0 ng/µl
2,4–xylidine	1.5 ng/µl
3,4–xylidine	3.µg/µl

Peak Identity:

1. m–xylene
2. 2,6–xylidine
3. 2,5–xylidine
4. 2,4–xylidine
5. 3,4–xylidine

Retention time (min)

Figure 4. Separation of xylidines by LSC. Used with the permission of the Liquid Chromatography Division of H. Reeve Angel & Co., Clifton, N.J.

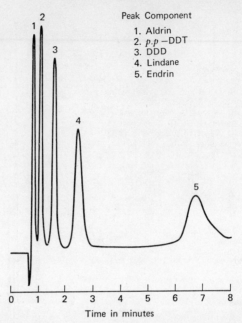

Peak Component

1. Aldrin
2. *p.p* —DDT
3. DDD
4. Lindane
5. Endrin

Figure 5. Liquid chromatogram of insecticides. Column: 50 cm × 2.3 mm inside diameter, containing Corasil II. Flow: 1.5 ml/min. Solvent: *n*-hexane. Pressure: 280 to 320 psi. Sample size: 5 µl. Reprinted from K. J. Bombaugh, R. F. Levangie, R. N. King, and L. Abrahams, *J. Chromatogr. Sci.* **8**, 657 (1970) by courtesy of the Preston Technical Abstracts Co.

The additional recommendations for LLC are (1) the thickness of the stationary liquid phase, d_f, should be small and (2) the diffusion coefficient for the solute in the stationary phase, D_s, should be large. The latter is promoted by liquids of low viscosity.

Stationary and Mobile Phases

Since both phases in LLC are liquids, they have particular properties and relationships that make it necessary to discuss them together. Usually the stationary liquid is chosen to have a polarity or functionality like that of the sample. The mobile phase must be immiscible with the stationary phase; consequently the two phases are quite dissimilar and of necessity are close to the extremes in polarity; that is, either the stationary phase is quite polar and the mobile phase is nonpolar or vice versa. Since the former condition is most common, it has become known as "normal" LLC and the opposite as "reverse phase" LLC (*18*). Even though the term "reverse phase" is widely

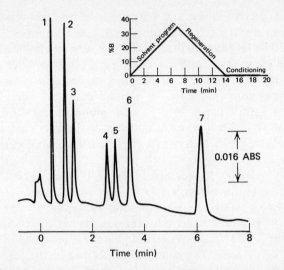

Figure 6. Solvent-programmed separation of antioxidants on 5-μm alumina. Column: 5-μm LiChrosorb Alox T; dimensions: 15 cm × 2.4 mm. Mobile phase: solvent A, hexane; solvent B, methylene chloride; flow rate, 60 ml/hr. Sample size, 1 μl; sample concentration, 2 mg/ml of each. Detector, ultraviolet. Samples: 1, BHT; 2, CAO-14; 3, triphenyl phosphate; 4, antioxidant 754; 5, BHA; 6, Goodrite, 7, Santowhite powder. Reprinted with permission from R. E. Majors, *Anal. Chem.* **45**, 755 (1973). Copyright by the American Chemical Society.

Table 8. Some Liquid Phases for LLC

Stationary	Mobile
Binary	
1. Polyethylene glycol, such as Carbowax 400 or triethylene glycol	1. Hydrocarbon such as pentane, isooctane, or cyclopentane; or a hydrocarbon plus a small amount of more polar liquid such as chloroform
2. β,β'-Oxydipropionitrile	2. Same as 1
3. Dimethylsulfoxide	3. Isooctane or hexane
4. Ethylenediamine	4. Same as 3
5. Tri-*n*-octylamine	5. Aqueous acid
6. Water–Alcohol or glycol	6. Hexane–carbon tetrachloride
7. Hydrocarbons and polymers	7. Water–alcohol
Ternary	
1. Water–ethanol–isooctane	
2. Chloroform–cyclohexane–nitromethane	

used, it is not clearly defined and suggests something abnormal about the chromatographic system. It is misleading, and its use should be discontinued.

Some typical liquid pairs are listed in Table 8. The mobile phase needs to be saturated with the stationary phase in each case. Note the difference between ethylenediamine and tri-*n*-octylamine. When ethylenediamine is used with a hydrocarbon mobile liquid, it acts as a polar stationary liquid; tri-*n*-octylamine is often used with an aqueous acid mobile liquid, and it becomes protonated and acts more like an ion-exchange stationary liquid. (The latter is classed as a reverse-phase system and illustrates the confusion this term can cause.)

Two ternary systems are included in Table 8. When properly chosen, a ternary system will form a large number of two-phase systems having a range of polarities. Either phase can be used as the stationary one and the other as the mobile one. The phase diagram for the water–ethanol–isooctane system is shown in Figure 7 (*15*). Thirteen pairs of phases are indicated by the pairs of points joined by tie lines. The compositions of two of the pairs, 1 and 13, which represent the extremes shown in Figure 7, are given in Table 9.

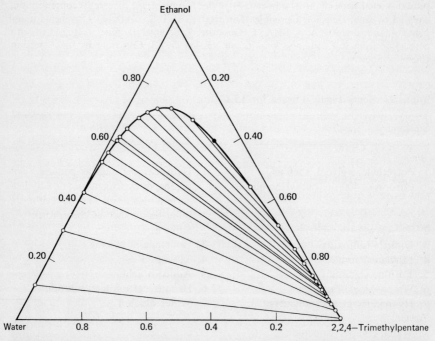

Figure 7. Phase diagram for the ternary system water, ethanol, isooctane (2,2,4-trimethylpentane). The solid circle indicates the plait point. Reprinted from J. K. Huber, *J. Chromatogr. Sci.* **9**, 72 (1971) by courtesy of the Preston Technical Abstracts Co.

Table 9. Compositions of Two of the Coexistent Pairs of Phases from Figure 7 for the System Water–Ethanol–Isooctane[a]

		Mole %		
Example		Water	Ethanol	Isooctane
1 {	Polar	88.7	11.3	<0.1
	Nonpolar	<0.1	0.9	99.1
13 {	Polar	11.3	66.6	22.1
	Nonpolar	5.2	44.0	50.8

[a] From Huber (15).

Example 1 has nearly pure water for one phase and nearly pure isooctane for the other. Example 13, on the other hand, is typical of a pair of phases that are nearly the same; if the two compositions become more alike, they will form only one, homogeneous, phase (plait point). Thus the range in polarity for the stationary phase can vary from the polar phase in example 1, to the polar phase in example 13, to the nonpolar phase in example 1, which represents a very wide range of possibilities. Obviously the mobile phase in each case is the other member of the pair; hence the range in polarity is equally large.

A pair of phases like example 13 has some limitations, however. Such a chromatographic system does not provide much selectivity; the phases are too similar in polarity. Second, it becomes difficult to keep the stationary phase immobile if it is very similar in composition to the mobile phase. It tends to be displaced by the mobile phase, and the system is unstable. This mechanical stability is largely determined by the solid support.

As in GLC, the stationary liquid phase is held immobile on an inert solid support. Both the porous materials (diatomaceous earth and silica) and porous layered materials can be used. The latter have smaller surface areas and cannot hold as much liquid (0.5 to 3%) as the porous supports (up to 30%). In either case, however, the surface of the solid support must have some attraction for the stationary liquid in order to hold it against the mechanical shearing forces of the liquid–liquid interface. Since the common supports have polar groups (e.g., hydroxy) on their surfaces, they are best suited for use with polar stationary liquids. As already noted, polar stationary phases are most common. Thus it can be seen why it is difficult to keep the nonpolar phase in example 13 from displacing the polar one: both have about the same attraction for the solid support. By the same reasoning, the solid support used for a nonpolar stationary phase should be nonpolar. The supports commonly used are Teflon and thoroughly deactivated (silanized) diatomaceous earth, silica, or carbon.

Chapter 13 contains further information about column packings and packing procedures. With respect to LLC specifically, it should be noted that (1) a precolumn is advisable to ensure that the mobile phase is saturated with stationary phase and that (2) the highly touted balanced density slurry method cannot be used for packing LLC columns. The stationary liquid phase can be loaded on the column in situ, however (*11, 19*).

Bonded Phases. When the stationary phase is chemically bonded to the support, some of the undesirable properties just mentioned are removed. The phase remains immobile and does not bleed; a precolumn is not required; any packing method can be used; and temperature control is less critical. In addition, a wider range of mobile phases can be used with a given stationary phase since the solubility of the stationary phase in the mobile phase is immaterial. Since bonded phases are also used in GC, additional information is included in Chapters 12 and 13.

Some differences have been reported between conventional liquid phases and bonded phases, but only one unified study has been reported to date (*20*). In that study polyethylene glycol (Carbowax) phases were compared at the same weight percent. In general, the two types of packings were found to be similar (comparing partition ratios k'), but the bonded phases were more efficient (larger number of plates). Overall, bonded supports have advantages that should result in their increasing popularity. One major limitation is the necessity to use small samples. The permanently bonded phases have low capacities.

Applications and Discussion

The LLC technique can be applied to a wide variety of sample types since the phases can be quite variable in polarity. This is especially true of bonded phases. Figures 8 and 9 show a bonded phase being used with a nonpolar mobile phase in one case (Fig. 8) and a polar mobile phase in the other (Fig. 9) (*21*). Another application more suited to LLC than LSC is the separation of members of a homologous series. Figure 10 shows the separation of a group of oligomers that comprise a commercial surfactant (*22*). The numbers on the peaks represent the number of ethylene oxide units.

On the other hand, gradient elution is not amenable to LLC and thus cannot be used to space evenly the members of a homologous series the way it is done in GLC with programmed temperature. However, gradient elution is possible with bonded phases, and this type of system is becoming very popular.

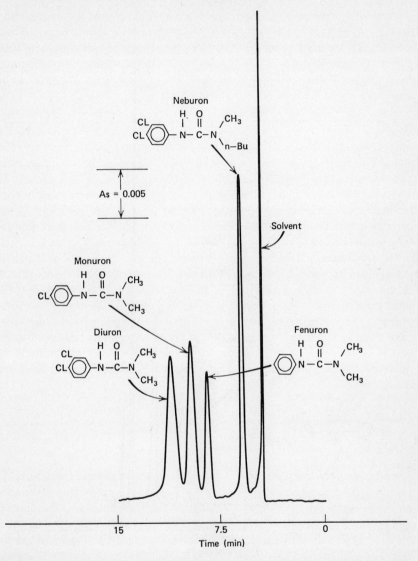

Figure 8. Use of a nonpolar mobile phase in the separation of ureas on ETH-Permaphase; Conditions: column, 1 m × 2.1 mm, packed with ETH-Permaphase; sample, 1.5 μl of 0.25 mg/ml each in methanol; mobile phase, 1 % dioxane in hexane; flow rate, 1.0 ml/min; temperature, 27°C; column pressure, 340 psig; ultraviolet detector. Reprinted with permission from J. J. Kirkland, *Anal. Chem.* **43** [12], 36A (1971). Copyright by the American Chemical Society.

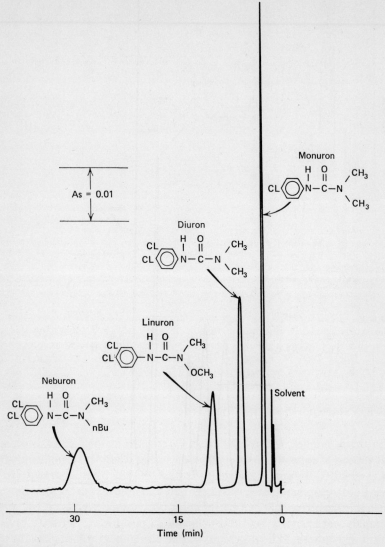

Figure 9. Use of a polar mobile phase in the separation of ureas on ETH-Permaphase. Conditions the same as in Figure 8 except for the following: sample, 4 μl of 0.25 mg/ml of each in methanol; mobile phase, 35% methanol in water; column pressure, 860 psig; temperature, 50°C. Reprinted with permission from J. J. Kirkland, *Anal. Chem.* **43** [12], 36A (1971). Copyright by the American Chemical Society.

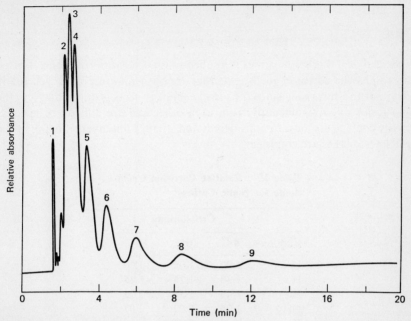

Figure 10. LLC Separation of oligomers that comprise a commercial surfactant. The numbers on the peaks represent the number of ethylene oxide units. Conditions: stationary phase, PEG-400; mobile phase, isooctane; $P_i = 42$ atm. Reprinted with permission from J. F. K. Huber, F. F. M. Kolder, and J. M. Miller, *Anal Chem.* **44,** 105 (1972). Copyright by the American Chemical Society.

ION-EXCHANGE CHROMATOGRAPHY

As the name implies, IEC is used primarily to separate ions, and this neccessitates the use of aqueous solutions. In these respects it differs significantly from LSC and LLC. In fact, the development of IEC has been separate from the other LC developments. IEC was developed and used mainly by inorganic and biochemists whereas LSC and LLC (and GLC) were developed by organic and analytical chemists, and this accounts in part for the separate development. Furthermore, IEC was the first liquid chromatographic technique to be automated (*23*); "amino acid analyzers" were commercially available long before the development of modern LC instruments. Only recently have the two separate approaches to instrumentation been merged.

Chapter 6 contains an introduction to the ion-exchange process and its application to chromatography as well as a list of general references. Kirkland's book (*24*) contains two chapters on IEC: a general one by Scott and one on the analysis of nucleic acids by Gere.

Theory

When classical thermodynamics is applied to ion-exchange reactions, it often fails to provide partition coefficients that are borne out in practice. Actually this is not too surprising since the ion-exchange process is quite complex, the solid resins differ significantly from each other and are difficult to specify, activity coefficients must be considered, and so on. Consequently many of the principles of IEC are empirical.

Table 10. Relative Partition Coefficients for Some Cations[a]

	Crosslinking		
Cation	4%	8%	10%
Li	1.00	1.00	1.00
H	1.30	1.26	1.45
Na	1.49	1.88	2.23
NH$_4$	1.75	2.22	3.07
K	2.09	2.63	4.15
Rb	2.22	2.89	4.19
Cs	2.37	2.91	4.15
Ag	4.00	7.36	19.4
Tl	5.20	9.66	22.2

[a] Source: Mallinckrodt Chemical Co.

The rules of selectivity toward metal ions are typical examples. For a sulfonic acid type of resin, the selectivities of monovalent cations are given in Table 10. Crosslinking refers to the resin porosity and will be discussed later. Clearly the resin with 10% crosslinking provides more selectivity in most cases, but there are exceptions like rubidium and cesium for which the selectivity is reversed.

For divalent cations the order is $UO_2 < Mg < Zn < Co < Cu < Cd < Ni < Ca < Sr < Pb < Ba$. Qualitatively these relative selectivities follow the theory of Eisenman, which has been summarized by Walton (*25*); Walton's chapter is recommended for further reading.

Kinetics. As with the other LC processes, the slow diffusion in the mobile liquid is a source of zone broadening. However, in relatively concentrated aqueous solutions (about 0.1 *M*) diffusion within the resin becomes the rate-determining step. In fact, diffusion coefficients for ions in aqueous solution *inside* an ion-exchange resin are 10 to 100 times slower than normal. Rapid exchange is favored by high temperature, small ions, low percentage cross-linking, as well as small resin-particle size (or pellicular design).

Types of Reaction. Care was taken in Chapter 6 to distinguish between chemical reactions and exchanges based on coulombic attraction. Although the division is not clear, another attempt to classify ion-exchange processes is made in Table 11.

Table 11. Classifications of Separations by Ion-Exchange Chromatography

Type	Example
1. Simple ion exchange:	
a. Cationic	Separation of Li^+, Na^+, K^+ on a cation resin
b. Anionic	Separation of halides on an anion resin
2. Competing equilibrium:	
a. Simple	Separation of carboxylate anions at pH 8 on an anion resin
b. Gradient elution	Separation of amino acids on cation resin with increasing pH
	Separation of metals on an anion resin by forming chlorocomplexes and decreasing HCl concentration
3. Mixed effects	Separation of analgesics on cation resin at pH 6.9
	Separation of sulfoxides on cation resin using nonaqueous solvents
4. Ligand exchange	Separation of amino acids on cation resin in the presence of Zn^{2+}

The processes referred to as simple (type 1) are those one would expect based on the relative partition coefficients as shown in Table 10. These separations are not very selective and not very common. Selectivity can be improved by incorporating a secondary equilibrium into the system, as in type 2. In type 2a the example shows a separation based on the relative strengths of the carboxylic acids and hence their respective degrees of ionization at pH 8. The gradient-elution separations (type 2b) have been classed separately since the change in mobile liquid can drastically alter the ionic nature of a solute. For example, an amino acid can be converted from a cation to a neutral molecule; or the chlorocomplexes can be changed from anions to cations, as described in detail in Chapter 6. This type of separation is very popular.

Type 3 is called mixed effects because these separations can involve some or all of the following: ion exchange, coulombic repulsion, adsorption, complexation, absorption, and molecular size sieving. Sometimes nonaqueous solvents are used and sometimes mixed aqueous/nonaqueous. The mechanisms are very complex, and no further clarification will be attempted.

In ligand exchange (type 4) a cation resin is used and equilibrated with a metal ion that is present in the mobile phase and is capable of complexing with the sample (and possibly other anions in the mobile phase). Thus the affinity of the sample for the resin depends on the strength of its metal complex. A number of equilibria must be considered. A recent separation of amphetamines has been reported (*26*) by this technique, and references to earlier separations can be found in this paper.

Stationary Phase

The most common synthetic ion-exchange resins are copolymers of styrene and divinylbenzene. The two-dimensional representation of the polymer is shown in Figure 11. The crosslinks between the carbon chains provide rigidity, and

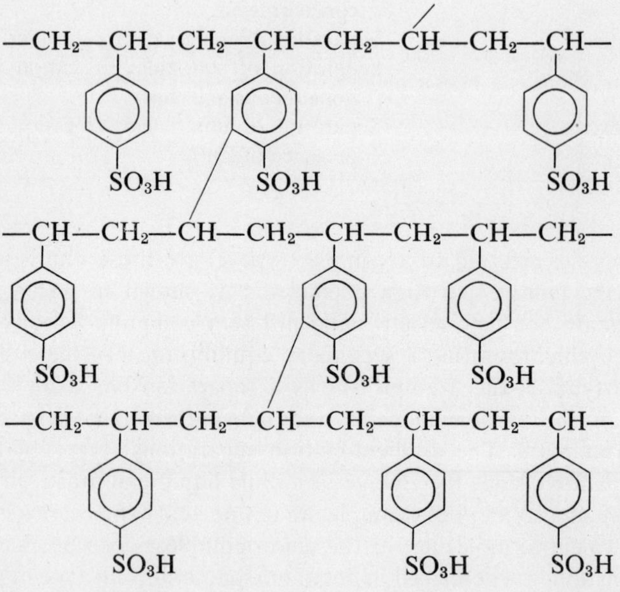

Figure 11. Chemical structure of crosslinked sulfonated polystyrene.

Table 12.   Common Types of Ion-Exchange Resins

Type	Cation; Strongly Acidic	Cation; Weakly Acidic	Anion; Strongly Basic	Anion; Weakly Basic
Functional group	$-SO_3^-H^+$	$-CO_2^-H^+$	$\begin{array}{c}CH_3\\ \mid \\ -N^+-CH_3\ Cl^-\\ \mid \\ CH_3\end{array}$	$\begin{array}{c}R\\ \mid \\ -N^+-H\ Cl^-\\ \mid \\ R\end{array}$
Trade name:				
Dowex	50W	—	1	3
Duolite	C-20	CC-3	A-101	A-2
Amberlite	IR-120	IRC-50	IRA-400	IR-45
Permutit	Q-100	Q-210	S-100	S-300

the number of crosslinks depends on the amount of divinylbenzene in the reaction mixture. The extent of crosslinking is commonly 4 to 16%. Highly crosslinked resins are harder, more brittle, more impervious, swell less, and are more selective. Also shown in the figure are sulfonic acid functional groups. This is a type of ion-exchange site that is formed on the polymer resin. Since the acid group has a labile proton, a cation, it is a cation-exchange resin. Anion types are also possible (the four common types are listed in Table 12). In addition, there is a chelating resin with the functional group

$$-CH_2-N\Big\langle {{CH_2-CO_2^-} \atop {CH_2-CO_2^-}}$$

Resins based on this polymer swell when they are wetted and exhibit a porous-gel structure. They are also soft and can be compressed under high pressure. Therefore, for use in high-pressure LC, at least 4% crosslinking is necessary, but too much crosslinking will decrease the pore size and permeability, and must also be avoided. We noted earlier that porous layer-type or pellicular ion-exchange resins are commercially available also. Still a third type is a resin with noncollapsible large pores, known as a macroreticular resin. It has a high porosity and a large internal surface area, which is desirable. It can be used with nonaqueous solvents since water is not needed to cause swelling.

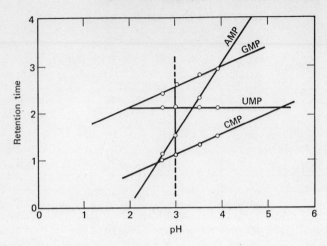

Figure 12.   Effect of pH on retention time. © 1972 Applied Science Laboratories, Inc.

Mobile Phase

The mobile phase is an aqueous solution, usually buffered and at a specific ionic strength. Large variations in retention behavior can be caused by varying ionic strength and pH as shown in Figure 12 for some nucleotides (27). In gradient elution either or both of these variables can be changed.

Applications and Discussion

A number of applications were listed in Table 11. Inorganic separations of metals are commonly done by the formation of anionic complexes as described. The data on partition coefficients are summarized in periodic tables like that in Figure 13.

Most of the recent applications have dealt with the separation of bio-chemicals. One of the triumphs of IEC was the separation of amino acids (28). Two of the early workers, Moore and Stein, received the Nobel prize in 1972 for their investigations of amino acids in which IEC played a major role. Now faster separations are being sought, and Figure 14 shows a 5-hour analysis on a single column.

Equally successful has been the separation of nucleic acids and their derivatives. It is anticipated that direct chromatographic analysis of body

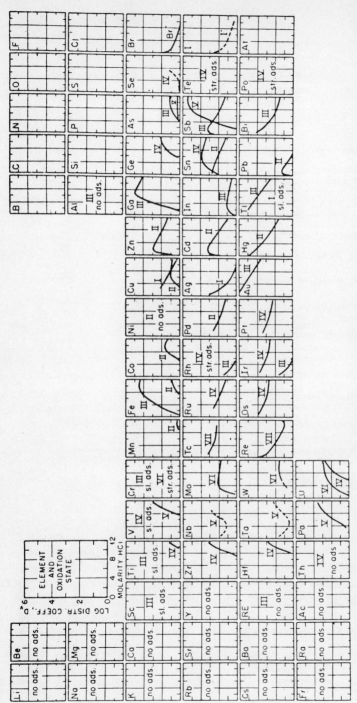

Figure 13. Absorption of the elements by Dowex 1-X10 from hydrochloric acid. Abbreviations: no ads., no adsorption; sl. ads., slight adsorption; str. ads., strong adsorption.

255

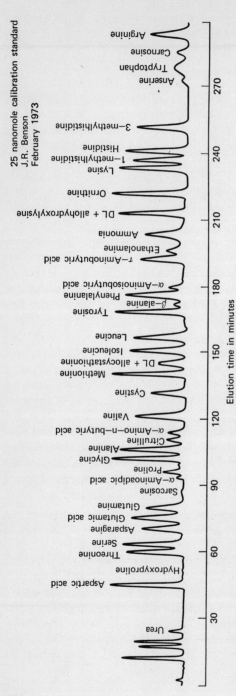

Figure 14. Fast separation of amino acids by IEC. Courtesy of the Durrum Chemical Corp.

fluids may soon permit the diagnosis of human diseases. Some pharma-
ceutical applications have been summarized (29).

GEL-PERMEATION CHROMATOGRAPHY

Gel-permeation chromatography (GPC) is one of the names given to the LC
method in which separations are based on a sieving process. In Chapter 5 it
was classified as a mechanical method of separation with some of the same
properties as molecular sieve chromatography in GC. This method is also
called gel-filtration, restricted-diffusion, and exclusion chromatography. The
most common names are GPC and exclusion chromatography; we shall use
the former.

The name "gel-permeation chromatography" was first used in 1962 among
polymer chemists who were separating organic polymers on polystyrene gels
in nonaqueous solutions. On the other hand, the name "gel-filtration
chromatography," described in 1959, was the name given to the separation of
biochemical polymers in aqueous solutions using dextran gels. Generally
speaking, these two techniques developed independently, even though the
processes are virtually the same. A single name should be used to describe
them, but the question is which name.

The term "exclusion chromatography" originates from work with ion-
exchange resins, which are very similar to some of the stationary phases used
in GPC. The exclusion of ions from an ion-exchange resin can be explained by
the Donnan membrane theory presented in Chapter 4 if one assumes that an
ion-exchange resin acts like a membrane separating the interior of the resin
from the bulk solution. If a strong cation resin is in the sodium form, for
example, sodium ions in the eluent will be excluded from the resin and pass
through the column unretained. On the other hand, a nonionic molecule like
glycol will distribute itself between the bulk solution and the interior of the
resin if the resin pores are large enough for it to enter. The diffusion of glycol
into and out of the resin will retard its migration through a column, and it will
elute after sodium chloride. Thus a mixture of sodium chloride and glycol can
easily be separated on a cation resin from which the sodium chloride is
excluded and eluted first.

Some molecules would be too large to enter the pores of the resin and they
too would be excluded and elute with the sodium chloride. Typically, sodium
chloride and sugar could not be separated this way. But a series of nonionic
molecules can be separated from each other as a result of selective sieving and
probably some adsorption. One classical separation is of sucrose, glycerol,
triethylene glycol, and phenol. This separation is certainly close to GPC, only
GPC resins do not have ionic sites like IE resins do. As a comprehensive
name, "exclusion chromatography" seems appropriate.

Two monographs have been written on GPC (*30, 31*) as well as several recent comprehensive papers, including a critical evaluation (*32*), LC on Sephadex (*33*), theory and mechanics (*34*), principles (*35*), a two-part series in the instrumentation section of the *Journal of Chemical Education* (*36*), and the practice of GPC (*37*).

Theory

We have stated that the principle on which separations are effected is a size-exclusion, or sieving, phenomenon. Before we begin to examine some of the characteristics of this process, it should be noted that this is only one of three proposed mechanisms. Bly (*38*) has classified them as follows:

1. Steric exclusion
2. Restricted diffusion
3. Thermodynamic considerations

Obviously there is an argument to be made for each. However, we shall restrict ourselves to steric exclusion, which is the preference expressed by Bly.

In this model the stationary phase, or gel, is a porous solid with pores whose sizes can be considered to be within well-defined limits. The sample is composed of molecules that vary in size and are in the range of the pore sizes in the gel. If a given solute is larger than the largest pores, it will be excluded and must spend all of its time in the mobile phase. Therefore it is eluted first. A slightly smaller molecule can enter a few of the pores, thus removing it from the flowing stream for a short time, so it elutes second. For the smaller and smaller molecules of the sample, longer times are spent inside the gel pores, and their respective retention times get longer and longer. There is a size of molecule, however, that can penetrate all the pores because it is smaller than the smallest pore, so it will have the longest retention time possible. A still smaller molecule cannot spend any more time in the column, so the limit of retention has been reached.

There are definite limits to the molecular sizes that can be separated on a given gel. The upper limit is a large molecule that is totally excluded, and the lower limit is a small molecule that totally permeates the resin. This effect is shown in Figure 15. Note that, unlike most chromatographic techniques, in this one the largest solutes are eluted first, followed by the smaller ones.

The classical chromatographic equation can be applied to GPC:

$$V_R^0 = V_M^0 + K_p V_S \tag{4}*$$

* The following special symbols have been used in the GPC literature to emphasize the special meaning of K_p and V_S, but it seems preferable here to emphasize the similarities, so they will not be used:

$$V_e = V_0 + K V_i$$

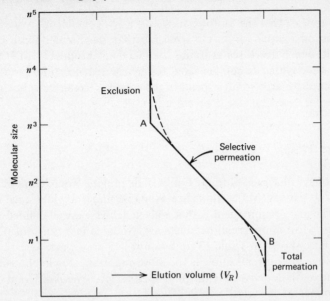

Figure 15. Illustration of permeation and exclusion limits in GPC. From K. M. Bombaugh, W. A. Dark, and R. F. Levangie, *Separ. Sci.* **3**, 375 (1968).

V_R^0 and V_M^0 have the usual meanings of total retention volume and mobile phase volume, respectively; V_S is considered to be the volume of the pores that are filled with stagnant mobile phase; K_p is a partition coefficient based on steric principles and not partitioning. In effect, K_p can vary from zero (total exclusion) to unity (total permeation). This is a severe limitation compared to other techniques. It means that all of the solutes to be separated must be eluted between a retention volume of V_M^0 and a retention volume of $(V_M^0 + V_S)$. The number of peaks that can be resolved in such a small range is small, and Giddings has made some approximations that indicate that the maximum number of peaks is between 10 and 20 (*39*).

In actual practice, partition coefficients greater than unity may be found because it is highly probable that some adsorption will occur for the retained solutes. This will depend on both the sample and the gel.

Rate Equation. Another characteristic of GPC separations that is not evinced in the other techniques is the constancy of the peak widths. To put it another way, in GPC the longer retained peaks have much smaller plate heights H than the earlier peaks. The rate equation should provide the answer.

The rate equation for LC was discussed in Chapter 13, and studies of axial diffusion in GPC have verified its validity for the mobile-phase terms (*40*).

Hence it appears that the uniform peak width is a coincidence resulting from compensating factors (37). It is possible that the mobile-phase diffusion coefficient, which decreases as solute size increases, produces band spreading for large solutes that is compensated for by the reduced time (and reduced band spreading) large solutes spend in the column. Recall that the "normal" elution order based on size is reversed in GPC.

Stationary Phases

The stationary phases used in GPC are solids ranging from synthetic organic polymers to glasses. Many kinds are available under a wide range of trade names. The main requirement is that a given gel have a well-defined range of pore sizes. Most commercial products are available in a range of pore sizes such as those shown in Table 13 for a controlled-pore glass. Selection of the proper range for a given sample is the most important step in choosing a gel. If one range will not cover the range of solute sizes, several different columns can be connected in series.

Other characteristics of the stationary phase are (1) its hardness or rigidity—some of the soft gels cannot be used at high pressure; (2) its compatibility with the solvent—the gel must be wetted by the mobile liquid to be used and for some gels the proper solvent is required to cause the gel to swell and open its pores; and (3) its surface activity—some gels will tend to adsorb some samples.

Table 13.   Specifications of a Typical Gel—Controlled Pore Glass, CPG-10[a]

Type (Average Pore Diameter, Å)	Range of Solute Diameters (Å)	Range of Molecular Weight (Polystyrene in Tetrahydrofuran)	Surface Area (m²/g)
75	80–35	1×10^4–2×10^3	340
120	120–45	3×10^4–3×10^3	210
170	170–55	6×10^4–6×10^3	150
240	240–65	2×10^5–9×10^3	110
350	350–80	3×10^5–1×10^4	75
500	500–90	6×10^5–2×10^4	50
700	700–120	1×10^6–3×10^4	36
1000	1000–150	3×10^6–4×10^4	25
1400	1400–180	5×10^6–6×10^4	18
2000	2000–200	2×10^7–9×10^4	13

[a] Source: Electro-Nucleonics, Inc.

Table 14. Types of Gel for GPC

Type and Trade Name	Composition	Solvents Used
1. Soft gels:		
Sephadex G	Crosslinked polydextran	Water
Sepharose B	Agarose gel	Water
Bio–Gel P	Polyacrylamide	Water
Bio–Gel A	Agarose gel	Water
Bio–Beads S	Styrene–divinylbenzene	Organic
2. Semirigid gels:		
Styragel	Styrene–divinylbenzene	Organic[a]
EM gel type OR	Crosslinked polyvinylacetate	Organic
Poragel	Styrene–divinylbenzene	Organic[a]
3. Rigid gels:		
Porasil (Spherosil)	Silica	Organic and water
EM gel type SI	Silica	Organic and water
CPG-10	Glass	Organic and water
Bio-glass	Glass	Organic and water

[a] Except acetone and alcohols.

An evaluation of available GPC packings has been published (*41*) but will undoubtedly become obsolete as new packings are prepared. A summary of available gels is given in Table 14.

Mobile Phase

There are few requirements the mobile liquid must meet. It should have a low viscosity like any LC solvent, and it should be a good solvent for the sample. Elevated temperatures can be used to improve sample solubility if it is a problem. If the gel swells in the solvent, it must be treated with solvent and allowed to swell before packing. The only other consideration is its potential deactivating effect on the gel surface if the gel is one that could adsorb the sample. In general, polar solvents are selected to prevent adsorption. Common solvents used in GPC are water, chloroform, and tetrahydrofuran.

Applications and Discussion

Gel-permeation chromatography is probably the easiest LC technique to understand and apply. It is used mainly to separate large molecules (molecular weights 2000 to 2,000,000) such as biochemicals and synthetic polymers,

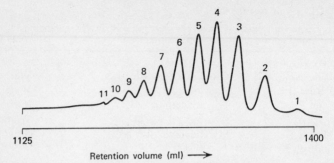

Figure 16. Separation of oligomers in a commercial surfactant (Triton X-45) by GPC. From K. M. Bombaugh, W. A. Dark, and R. F. Levangie, *Separ. Sci* **3**, 375 (1968).

although small molecules (molecular weights 200 to 400) can be separated with some gels (*42*). Cazes (*36*) lists nearly 100 *types* of compound that have been studied by GPC.

A typical separation of a surfactant is shown in Figure 16. It should be compared with the LLC separation of similar sample shown in Figure 10. By comparison, the GPC column and the separation time are much longer. In LLC the smallest molecules are eluted first, and in GPC the largest come off first.

Gel-permeation chromatography can be used to determine approximate molecular weights. Figure 17 shows a typical calibration curve for hydro-

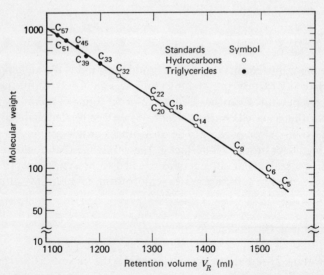

Figure 17. Typical molecular-weight calibration curve. From K. M. Bombaugh, W. A. Dark, and R. F. Levangie, *Separ. Sci.* **3**, 375 (1968).

carbons and triglycerides. In fact GPC is widely used to determine the molecular weight range of synthetic polymers. Only one peak is obtained since it is neither necessary nor desirable to achieve a better separation. An elementary discussion of the relationship between peak shape and polymer molecular weight can be found in Cazes's paper (*43*).

REFERENCES

1. L. R. Snyder, in *Advances in Analytical Chemistry and Instrumentation*, Vol. 3, C. N. Reilley (ed.), Wiley–Interscience, New York, 1964, p. 251.
2. C. H. Giles and I. A. Easton, in *Advances in Chromatography*, Vol. 3, J. C. Giddings and R. A. Keller (eds.), Dekker, New York, 1966, p. 67.
3. L. R. Snyder, *Principles of Adsorption Chromatography*, Dekker, New York, 1968.
4. L. R. Snyder, in *Modern Practice of Liquid Chromatography*, J. J. Kirkland (ed.), Wiley–Interscience, New York, 1971, Chapters 4 and 6.
5. R. W. Yost and R. D. Conlon, *Chromatography Newsletter* (Perkin–Elmer Corp.) 1[1], 5 (1972).
6. L. R. Snyder, *J. Chromatogr. Sci.* **8**, 692 (1970).
7. L. R. Snyder and D. L. Saunders, *J. Chromatogr. Sci.* **7**, 195 (1969).
8. R. P. W. Scott and P. Kucera, *J. Chromatogr. Sci.* **11**, 83 (1973).
9. R. P. W. Scott and P. Kucera, *Anal. Chem.* **45**, 749 (1973).
10. K. J. Bombaugh, R. F. Levangie, R. N. King, and L. Abrahams, *J. Chromatogr. Sci.* **8**, 657 (1970).
11. R. E. Majors, *Anal. Chem.* **45**, 755 (1973).
12. A. J. P. Martin and R. L. M. Synge, *Biochem. J.* **35**, 91 (1941).
13. J. J. Kirkland, *Modern Practice of Liquid Chromatography*, Wiley–Interscience, New York, 1971, Chapter 5.
14. D. C. Locke, in *Advances in Chromatography*, Vol. 8, J. C. Giddings (ed.), Dekker, New York, 1969, p. 47.
15. J. F. K. Huber, *J. Chromatogr. Sci.* **9**, 72 (1971).
16. E. Soczewinski, in *Advances in Chromatography*, Vol. 5, J. C. Giddings and R. A. Keller (eds.), Dekker, New York, 1968, p. 3.
17. J. F. K. Huber, C. A. M. Meyers, and J. A. R. J. Hulsman, *Anal. Chem.* **44**, 111 (1972); J. F. K. Huber, E. T. Alderlieste, H. Harren, and H. Poppe, *Anal. Chem.* **45**, 1337 (1973).
18. E. Cerrai and G. Ghersini, in *Advances in Chromatography*, Vol. 9, J. C. Giddings and R. A. Keller (eds.), Dekker, New York, 1970, p. 3; U. A. Th. Brinkman and G. deVries, *J. Chem. Educ.* **49**, 244 (1972).
19. J. F. K. Huber, J. C. Kraak, and H. Veening, *Anal. Chem.* **44**, 1554 (1972).
20. D. F. Horgan, Jr., and J. N. Little, *J. Chromatogr. Sci.* **10**, 76 (1972).
21. J. J. Kirkland, *Anal. Chem.* **43** [12], 36A (1971).
22. J. F. K. Huber, F. F. M. Kolder, and J. M. Miller, *Anal. Chem.* **44**, 105 (1972).
23. D. H. Spackman, W. H. Stein, and S. Moore, *Anal. Chem.* **30**, 1190 (1958).

24. J. J. Kirkland (ed.), *Modern Practice of Liquid Chromatography*, Wiley–Interscience, New York, 1971, Chapter 8 by C. D. Scott and Chapter 12 by D. R. Gere.

25. H. F. Walton, in *Chromatography*, 2nd ed., E. Heftmann (ed.), Reinhold, New York, 1967, Chapter 13.

26. C. M. deHernandez and H. F. Walton, *Anal. Chem.* **44**, 890 (1972).

27. *Gas-Chrom Newsletter* **13** [6], 3 (1972), Applied Science Laboratories.

28. P. B. Hamilton, in *Advances in Chromatography*, Vol. 2, J. C. Giddings and R. A. Keller (eds.), Dekker, New York, 1966, p. 4.

29. J. B. Smith, J. A. Mollica, H. K. Goran, and I. M. Nunes, *Am. Lab.* **4** [5], 13 (1972).

30. H. Determann, *Gel Chromatography* 2nd ed., Springer–Verlag, New York, 1969.

31. J. F. Johnson and R. S. Porter (eds.), *Analytical Gel Permeation Chromatography*, Wiley–Interscience, New York, 1968.

32. R. L. Pecsok and D. Saunders, in *Separation Techniques in Chemistry and Biochemistry*, R. A. Keller (ed.), Dekker, New York, 1967, p. 81.

33. J. Sjovall, E. Nystrom, and E. Haahti, in *Advances in Chromatography*, Vol. 6, J. C. Giddings and R. A. Keller (eds.), Dekker, New York, 1968, p. 119.

34. K. H. Altgelt, in *Advances in Chromatography*, Vol. 7, J. C. Giddings and R. A. Keller (eds.), Dekker, New York, 1969, p. 3.

35. H. Determann, in *Advances in Chromatography*, Vol. 8, J. C. Giddings and R. A. Keller (eds.), Dekker, New York, 1970, p. 3.

36. J. Cazes, *J. Chem. Educ.* **47**, A461 and A505 (1970).

37. K. J. Bombaugh, in *Modern Practice of Liquid Chromatography*, J. J. Kirkland (ed.), Wiley–Interscience, New York, 1971, Chapter 7.

38. D. D. Bly, *Science* **168**, 527 (1970).

39. J. C. Giddings, *Anal. Chem.* **39**, 1027 (1967).

40. R. N. Kelley and F. W. Billmeyer, Jr., *Anal. Chem.* **41**, 874 (1969).

41. W. A. Dark and R. J. Limpert, *J. Chromatogr. Sci.* **11**, 114 (1973).

42. K. M. Bombaugh, W. A. Dark, and R. F. Levangie, *Separ. Sci.* **3**, 375 (1968).

43. J. Cazes, *J. Chem. Educ.* **43**, A567 and A625 (1966).

PLANE CHROMATOGRAPHY

There are two liquid chromatographic techniques in which the stationary bed is supported on a planar surface rather than in a column: paper chromatography (PC) and thin-layer chromatography (TLC). Because they have much in common, they are treated together in this chapter. Extensive work in plane chromatography dates from about 1944 for PC and 1956 for TLC. Historically many separation methods were first developed on PC and then adapted to TLC, so that in many cases TLC has displaced PC and is more widely used today. However, there is still considerable work being done by PC (*1*).

Actually there are many types of "layer" used in TLC, including cellulose, and many new kinds of paper, including glass "paper" and papers "loaded" with solids e.g., silica gel. Thus the differences between PC and TLC are being narrowed. In general, in this chapter, "paper" refers to the normal, cellulose variety, and "thin layer" refers to silica gel or a similar adsorbent.

There is some confusion over attempts to classify the techniques according to the principles on which the separations are based. Classically, the water content of chromatographic paper is considered to be large enough for water to be considered a stationary liquid phase, the separation mechanism being absorption. On the other hand, the silica gel used in TLC is dried before use; the stationary phase is considered to be the solid, and the separation mechanism is adsorption. However, it has been found that a dry TLC plate adsorbs more than half its equilibrium concentration of water from a room atmosphere of about 50% relative humidity within 3 minutes. Consequently, unless special precautions are taken, TLC plates exposed to the atmosphere for even a short time (for sample spotting) will contain a significant amount of water, and some absorption will probably be involved in its use. Conversely, the water content of paper is insufficient to deactivate completely its surface, so some surface adsorption and ion exchange are probably involved in its use. In practice these considerations are secondary; primary importance is attached to the repeatability of preparing paper or TLC plates. Since the actual situation is complex, it will not be considered further.

It should be recalled from Chapter 2 that the parameter commonly used to express solute behavior in plane chromatography is the retardation factor, R_F, not the retention volume (or retention time) as in the column method. This choice is a practical rather than a theoretical one since either parameter is satisfactory, and their relationship has been adequately discussed in Chapters 2 and 11.

THEORY

Since the theories of chromatography have been discussed, this section emphasizes the topics that are of special importance in the planar methods and the differences between planar and columnar methods.

Comparison with Column Methods

Mobile-Phase Flow. In plane chromatography the flow rate of the mobile liquid cannot be controlled as it can in the column methods. It is dependent on the surface tension, τ, and the viscosity, η, of the mobile liquid as well as the nature of the stationary bed. One model of the system is the interconnected capillary model shown in Figure 1. The bed is composed of capillaries of varying diameter, connected to each other by capillaries perpendicular to the direction of flow.

Initially the bed is dry; the solvent is applied at one end and enters the capillaries. The surface tension causes the solvent to be more strongly attracted to the small capillaries, which are filled first and in which the solvent front advances. As it does so, solvent is drawn into the capillaries that have the least resistance to flow—the large ones. Since all sizes are interconnected, bulk solvent flow is possible from the large capillaries to the small ones. The result

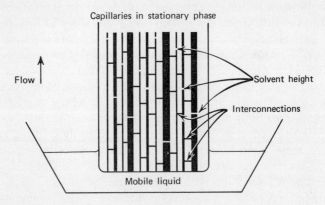

Figure 1. Interconnected capillary model of plane chromatography.

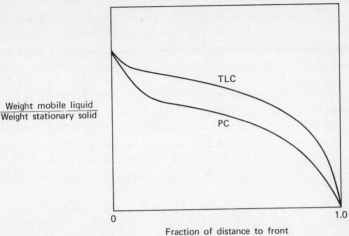

Figure 2. Variation in solvent concentration along direction of flow.

is that the solvent front is moving rapidly in the small capillaries, and the bulk solvent is moving more slowly in the large capillaries. The more heterogeneous the capillary size, the more variation in flow. In general, paper is more heterogeneous than a silica gel bed, so the flow variation is greatest in paper.

On the basis of this model, one would expect that the quantity of solvent on the bed would vary from the front to the reservoir. Experimental verification of this phenomenon is shown in Figure 2. As expected, the gradient is more pronounced in PC (2) than in TLC (3).

The distance the solvent front moves, s_F, has been shown to be

$$s_F = (kt)^{1/2} \tag{1}$$

where t is the time and k is a proportionality constant that is directly proportional to the surface tension and inversely proportional to the viscosity.

$$k \propto \frac{\tau}{\eta} \tag{2}$$

Furthermore, the velocity of the solvent front (u_F) is

$$u_F = \frac{ds_F}{dt} = \frac{k}{2s_F} \tag{3}$$

Thus we can conclude that (1) the velocity of the solvent front is proportional to the ratio k of surface tension to viscosity of the solvent, and (2) the velocity is inversely proportional to the distance the front has moved, s_F, so its velocity decreases during a run. The flow is not constant in distance or in time, and this is a major obstacle to a theoretical treatment. Furthermore, the flow cannot be controlled. In column methods the flow was always assumed to be constant and could be easily controlled.

There are two other undesirable characteristics of the flow in planar systems. One arises from the fact that the solvent is moving on a dry bed, so some heat will be given off when it becomes wet. Consequently it is expected that a small temperature differential will exist between the front and the rest of the bed. A related phenomenon is the problem associated with evaporation of solvent from the bed (4). To prevent evaporation, the system is usually closed and kept saturated. The temperature is seldom controlled, however, so a slight decrease in temperature will result in undesirable condensation.

Stationary Phase.   The composition of the stationary phase, especially the water content, is difficult to control. Furthermore, the composition of the solid can vary considerably, especially in the case of paper. As a consequence of this stationary-phase heterogeneity (and the lack of constant temperature), thermodynamic principles cannot be applied to the system with the same precision as in the column methods. It is understandable that the approach to plane chromatography has been largely empirical.

Rate Equation and Resolution

Theoretical considerations of band broadening are made difficult by the lack of a constant flow rate (mobile-phase velocity) since it is an important parameter in the rate equation. Nevertheless, studies have been made using the rate equation of Giddings (see Chapter 8) as the basis (3, 5, 6). The conclusions can be summarized:

1. Band broadening due to longitudinal molecular diffusion is significant in PC and greater than in TLC due to the larger tortuosity factor ψ of paper despite the faster flow rate in PC.
2. The greater variation in flow in PC makes it less efficient than TLC.
3. Mass transfer in the mobile phase is slower in PC than in TLC, probably due to the larger effective particle size; this makes PC less efficient than TLC.

Thoma and Perisho (7) have summarized the theory for planar chromatography and suggested guidelines for the selection of the best chromatographic conditions. Their main conclusion is that optimum resolution is obtained when R_F is about 0.25 and that a solvent should be chosen to achieve this performance. It is interesting to note that an R_F of 0.25 corresponds to a partition ratio k' of 3.0, which is approximately the value recommended for LC in columns.

In practice, other errors can be made that will produce band broadening, which is more important than the theory predicts; for example, the sample spot may be too large or distorted in shape, the bed may contain impurities that can cause streaking, or the sample components may undergo displacement chromatography when placed on the bed (ref. *8*, p. 21).

Other Equations

Martin Equation. In 1949 Martin described an approach toward a relationship between solute structure and free energy for chromatographic partitioning (*9*). The next year this was put in terms of the newly defined parameter R_M:

$$R_M = \log \left(\frac{1}{R_F - 1} \right) = \log k' \tag{4}$$

and since $k' = K_p/\beta$, R_M is a measure of the $\log K_p$ and the free energy (Chapter 3). By comparing R_M values of compounds that differ by a methylene group or a functional group, one can compile a list of R_M values for various groups. If they are additive (as free energies should be), the R_M value for any compound can be predicted by summing the values for the groups of which the molecule is composed. Hence the Martin equation is

$$(R_M)_T = \sum_i (R_M)_i \tag{5}$$

Applications to PC have been summarized (*10*), but some workers have noted that this concept does not always apply in TLC (*11*).

Snyder Equation. In the section on LSC in Chapter 14 the Snyder equation was presented:

$$\log K_p = \log V_a + E_a(S^0 - \epsilon^0 A_s) \tag{6}$$

Since both LSC and TLC involve adsorption from solution, this equation applies equally well to TLC. Note also that equation 6 can be written in terms of R_M:

$$R_M = \log \left(\frac{V_a}{\beta} \right) + E_a(S^0 - \epsilon^0 A_s) \tag{7}$$

For further discussion see reference *11*.

APPARATUS AND TECHNIQUES

Since most of the equipment and many of the procedures are the same for PC and TLC, they will be considered together. For more specific information on a particular apparatus or technique, the monographs listed in the bibliography can be consulted. The greatest difference between the techniques is the stationary phase—the paper or the thin layer itself. Extensive listings of commercially available papers, prepared piates, and flexible sheets and equipment have been published (*12*).

Stationary Phase

Paper. Although ordinary filter papers have been used, special papers are manufactured for PC. They are highly purified (freed from metals) and are manufactured under controlled conditions so that such properties as porosity, thickness, and arrangement of cellulose fibers are reproducible from batch to batch. A single paper, Whatman No. 1, can be used for most separations. The theoretical aspects related to the use of paper in chromatography have been summarized by Stewart (*13*).

As already noted, undried paper contains enough adsorbed water for PC to be classified as an absorption process. Other liquids can be applied to the paper to change its characteristics. For example, silicone oils, petroleum jelly, paraffin oil, and rubber latex are used as nonpolar liquid phases (so-called reversed phases). Special papers are also commercially available that contain adsorbents or ion-exchange resins or are specially treated (e.g., acylated) or are made of other fibers (e.g., glass, nylon).

Thin Layers. Silica gel is the solid most commonly used in TLC. Several types are available, depending on whether or not a binder and/or a fluorescent indicator is added. Other popular solids are alumina, cellulose, and poly-amide. The solid can be coated on glass, aluminum, or plastic sheets. The usual procedure is to coat silica gel on glass, but a wide variety of precoated plates and sheets are commercially available (ref. *14*, especially part I). As with paper, the solid layer can be coated with a liquid stationary phase. The thickness of the solid layer can vary from 150 to 500 μm with thicker layers (up to 2 mm) being used for preparative work. For fast routine scanning small microscope slides can be used (*15*). A review article has recently appeared on the stationary phase in TLC (*16*).

In plane chromatography a variety of shapes can be used for the bed in addition to the conventional rectangular shape. Some are circles, wedges, cylinders, rods, and even drums. This adds a degree of flexibility not possible in column chromatography.

Mobile Phase and Selectivity

The mobile phase is a liquid or a mixture of liquids. Although mixtures can be problematic due to demixing, preferential evaporation, and preferential adsorption on the layer, they are quite common. A typical mixture is butanol–acetic acid–water.

The choice of a mobile phase is determined by the nature of the stationary phase and generally follows the guidelines established in Chapter 14. For example, with a silica gel stationary phase the parameter used to select the mobile phase would be Snyder's ϵ^0 or Hildebrand's solubility parameter, and a deactivator may be added. If a liquid is used as the stationary phase, a mobile phase of opposite polarity and low solubility is chosen. Gradient techniques can be used in the mobile phase and in the stationary phase (17).

Development and Apparatus

The prepared plate or paper is developed in a closed, presaturated chamber using an ascending or descending mobile solvent flow. Four possibilities are shown in Figure 3. Whenever paper is used, it must be supported (as in Fig. 3a, b, and d); most thin layers are self-supporting.

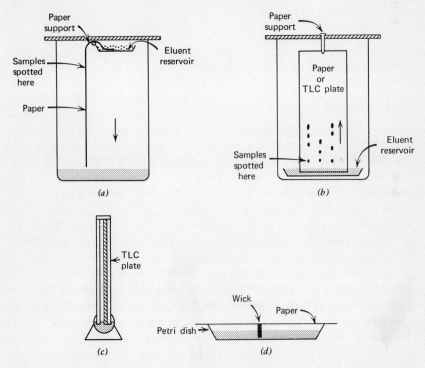

Figure 3. Developing chambers for plane chromatography: (a) descending—used with PC; (b) ascending—used with PC and TLC; (c) sandwich—used with TLC; (d) horizontal—used with PC.

The sample is applied in a volatile solvent without disturbing the surface of the stationary phase. Its area must be kept as small as possible, and the solvent is evaporated before development is begun. The sample is placed 1 to 2 cm from the edge of the plate that will be immersed in the mobile liquid. The actual distance can have a significant effect on R_F values, especially in PC with mixed solvents (ref. *18*, p. 161).

Normal development varies from about 15 minutes (for TLC) to several hours (for descending PC). In addition to the normal procedures some special techniques are the following:

1. Multiple development, in which the plate is developed several times when R_F values are small.

2. Programmed multiple development, in which the plate is automatically cycled through a number of developments (*19*).

3. Overrun, which is allowed to take place also to increase R_F values.

4. Two-dimensional development, in which a second solvent is run perpendicular to the first solvent flow on a square plate after the first development is completed and dried.

5. A special type of programmed-temperature TLC (*20*).

A great deal of more detailed information is available on PC (*8, 18, 21–23*) and TLC (*13, 24–27*) apparatus and techniques.

Visualization and Detection

After the development is completed and the plate (or paper) dried, the solute bands must be located unless they themselves are colored. Bobbitt (*28*) lists six universal spray reagents, including sulfuric acid and iodine (p. 86) and 89 other specific reagents (pp. 88–102) for use in TLC. Some of these, like sulfuric acid, cannot be used on paper, but most of the specific spray reagents can be used in both PC and TLC. Solutes can also be visualized under an ultraviolet lamp if they are fluorescent. If they are not, another procedure is to add a fluorescent material to the stationary phase when it is prepared. When the developed plate is viewed under ultraviolet radiation, the plate will fluoresce and the solute bands will not.

For qualitative analysis, the distances the solutes have moved from the origin are measured and compared to the distance the solvent front has moved by calculation of the R_F values. Alternatively, greater reproducibility can be obtained by comparing the solute distances to an internal standard. The R_M values can be calculated by equation 4. The colors of the bands themselves can sometimes be used for qualitative analysis also.

For quantitative analysis, the solute bands can be cut out of the paper or scraped off the plate, extracted, and determined using an appropriate instrument. Alternatively, the plate can be scanned with a densitometer, which produces a differential chromatogram like the ones from the column methods (29). This is very convenient and is becoming more popular, although column methods are still preferred for quantitative analysis. A semi-quantitative analysis for plane chromatography can be accomplished by measuring the areas of the solute spots.

COMPARISON OF THIN-LAYER AND PAPER CHROMATOGRAPHY

For the most part TLC is preferred over PC because it is faster. However, there are some specific analyses that are reported to be better on PC (1). The advantages of each technique are summarized in Table 1.

Table 1. Comparison of Some of the Advantages of TLC and PC

TLC	PC
1. Faster	1. Cheaper
2. Greater efficiency per unit time	2. Easier to quantitate (compared to
3. More versatile; greater number of	scraping plates)
stationary phases	3. Easier to use descending technique
4. Easier to use; better handling and	
visualization	
5. Slightly more sensitive	

COMPARISON OF LC SEPARATIONS IN COLUMNS AND ON PLANE SURFACES

In many respects, LC separations in columns and on plane surfaces are similar. However, only a few studies comparing these separations have actually been made.

In their work with steroids, Kabasakalian and Talmage (30) derived the following equation for comparing paper chromatography (p) and liquid–liquid chromatography in columns (c):

$$(R_M)_c = (R_M)_p + \log\left(\frac{\beta_p}{\beta_c}\right) \tag{8}$$

They were able to show that this relationship holds, and they obtained values for β_p/β_c graphically. With the latter ratio they could then predict column R_M (and R_F) values from PC R_M values. The general problem of determining optimum solvent systems for LLC and LLE from PC data has been discussed (31).

The concept of using simple TLC to find the best conditions for LSC in columns is attractive because of the low cost and ease of changing solvents. Since virtually the same silica products are now available for TLC and high-pressure LC in columns, the prospect looks bright,* The best evidence to date has been provided by Majors (*32*). His TLC and LSC separations of azo dyes are shown in Figure 4; the dyes are identified in Table 2. While the R_R values in Table 2 are only estimates, the two separations are clearly similar.

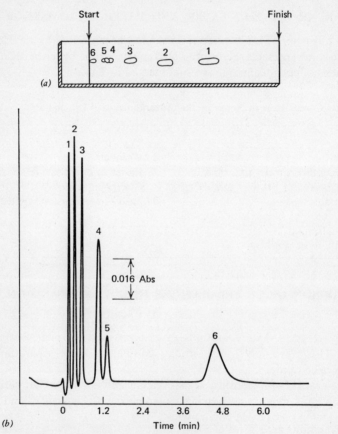

Figure 4. Separation of azo dyes by TLC (*a*) and LSC (*b*). In the TLC separation the plate was an E. Merck precoated TLC sheet, silica gel F-254; solvent, 10% CH_2Cl_2 in hexane; development time, 50 min. In the LSC separation the column was MicroPak SI-10; dimensions, 15 cm × 2.4 mm; mobile phase: 10% CH_2Cl_2 in hexane; flow rate, 132 ml/hr; pressure, 350 psi; sample size: 1 μl; sample concentration, 0.2 mg/ml each; ultraviolet detector. See Table 2 for a list of the compounds. Reprinted with permission from R. E. Majors, by *Anal. Chem.* **45**, 755 (1973). Copyright by the American Chemical Society.

* Note, however, that most TLC silicas contain a binder, and this may interfere.

Table 2. Dyes Separated in Figure 4

Number	Structure	TLC R_F	LSC $R_R{}^a$
1	(phenyl)—N=N—(phenyl)	0.69	0.55
2	(phenyl, Br)—N=N—(phenyl)—NEt$_2$	0.39	0.32
3	(phenyl)—N=N—(phenyl)—NEt$_2$	0.21	0.21
4	(phenyl, NO$_2$)—N=N—(phenyl)—NEt$_2$	0.098	0.12
5	O$_2$N—(phenyl)—N=N—(phenyl)—NEt$_2$	0.064	0.097
6	(phenyl)—N=N—(phenyl)—NH$_2$	0.015	0.029

a Estimated from Figure 4b.

Similar results are shown in Figure 5 for a series of oligophenylenes run by classical LC, high-pressure LC, and TLC (*33*). A densitometer tracing is given for the TLC separation from which compound "*a*" (benzene) is missing. Note the reversal in the order of the peaks (from the column methods) and the bands (from TLC). Of the three, high-pressure LC gives the best resolution in the least time, but TLC would seem to provide a satisfactory screening procedure.

Bombaugh (*34*) has provided another comparison, shown in Figure 6, but the TLC bands are not correlated with the column peaks. In this case the higher efficiency of LSC in columns is again evident.

Although the examples just given have indicated a fairly good correlation between the two types of chromatography, one should not assume that all published TLC or PC procedures can be converted to column operation. There are many problems in finding equivalent stationary phases, and many TLC (and PC) solvents cannot be used in columns due to high viscosities and detector incompatibility (e.g., ultraviolet absorption).

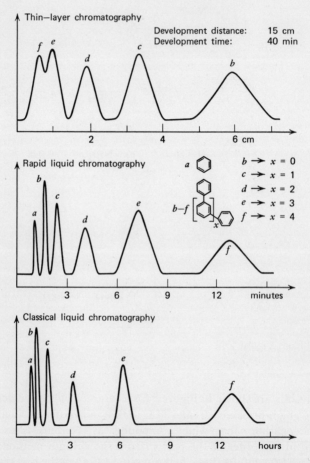

Figure 5. Comparison of the analysis of *m*-oligophenylenes by TLC, high-speed LC, and open-column LC. Sample was separated using E. Merck silica gel type 60 (40–63 μm) deactivated with 40% H_2O adsorbent and *n*-heptane mobile phase. From F. Eisenbeiss, *Chem. Zeit.* **95**, 237 (1971).

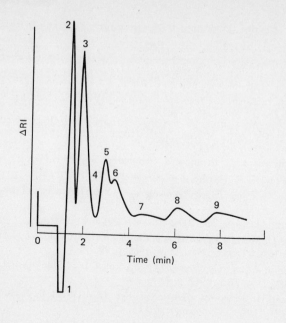

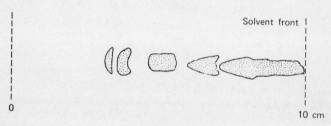

Figure 6. Comparison of high-performance column chromatography with TLC for separating polypropylene glycol (Ucon 50HB55). Top: column, 50 cm of Corasil II; carrier, n-hexane containing 5% isopropanol; flow rate, 0.9 ml/min; sample, 0.8 mg. in 20 μl solvent; complete separation, 9 min. Bottom: (reproduction of a thin-layer plate) developing solvent, n-hexane containing 20% ethanol; spot developer, iodine vapor; time, 40 min. Reprinted from J. J. Kirkland (ed.) *Modern Practice of Liquid Chromatography*. Copyright by John Wiley & Sons.

SUMMARY

Plane chromatography finds use as a simple, fast screening technique with a limited efficiency (about 300 plates). The range of applications is very wide, including inorganic ions and most classes of organic compounds. The bibliography can be consulted for details.

REFERENCES

1. G. Zweig and J. Sherma, *J. Chromatogr. Sci.* **11**, 279 (1973).

2. J. C. Giddings, G. H. Stewart, and A. L. Ruoff, *J. Chromatogr.* **3**, 239 (1960).

3. G. H. Stewart, in *Separation Techniques in Chemistry and Biochemistry*, R. A. Keller (ed.), Dekker, New York, 1967, p. 287.

4. G. H. Stewart and T. D. Gierke, *J. Chromatogr. Sci.* **8**, 129 (1970).

5. G. H. Stewart, *Separ. Sci.* **1**, 135 (1966).

6. C. L. DeLigny and A. G. Remijnse, *J. Chromatogr.* **35**, 257 (1968).

7. J. A. Thoma and C. R. Perisho, *Anal. Chem.* **39**, 745 (1967).

8. J. Sherma and G. Zweig. *Paper Chromatography*, Vol. II, Academic Press, New York, 1971.

9. A. J. P. Martin, *Biochem. Soc. Symp. (Cambridge)* **3**, 4 (1949).

10. J. Green and D. McHale, in *Advances in Chromatography*, Vol. 2, J. C. Giddings and R. A. Keller (eds.), Dekker, New York, 1966, p. 99.

11. L. R. Snyder, in *Advances in Chromatography*, Vol. 4, J. C. Giddings and R. A. Keller (eds.), Dekker, New York, 1967, p. 3.

12. K. Macek and H. Becoarova, *Chromatogr. Rev.* **15**, 1 (1971). R. J. Hurtubise, P. F. Lott, and J. R. Dias, *J. Chromatogr. Sci.* **11**, 476 (1973).

13. G. H. Stewart, in *Advances in Chromatography*, Vol. 1, J. C. Giddings and R. A. Keller (eds.), Dekker, New York, 1965, p. 93.

14. P. F. Lott and R. J. Hurtubise, *J. Chem. Educ.* **48**, A437, A481 (1971).

15. J. J. Peifer, *Mikrochim. Acta* **1962**, 529.

16. R. M. Scott, *J. Chromatogr. Sci.* **11**, 129 (1973).

17. A. Niederwieser and C. C. Honegger, in *Advances in Chromatography*, Vol. 2, J. C. Giddings and R. A. Keller (eds.), Dekker, New York, 1966, p. 123.

18. H. G. Cassidy, *Fundamentals of Chromatography*. Vol. X of *Technique of Organic Chemistry*, Wiley–Interscience, New York, 1957, Chapter VII.

19. J. A. Perry, K. W. Haag, and L. J. Glunz, *J. Chromatogr. Sci.* **11**, 447 (1973).

20. S. Turina and V. Jamnicki, *Anal. Chem.* **44**, 1892 (1972).

21. H. J. Pazdera and W. H. McMullen, in *Treatise on Analytical Chemistry*, Part I, Vol. 3, I. M. Kolthoff and P. J. Elving (eds.), Wiley–Interscience, New York, 1961, Chapter 36.

22. V. C. Weaver, in *Advances in Chromatography*, Vol. 7, J. C. Giddings and R. A. Keller (eds.), Dekker, New York, 1968, p. 87.

23. K. Macek, in *Chromatography*, 2nd ed., E. Heftmann (ed.), Reinhold, New York, 1967, Chapter 7.

24. I. Smith, in *Advances in Chromatography*, Vol. 1, J. C. Giddings and R. A. Keller (eds.), Dekker, New York, 1965, p. 61.

25. E. V. Truter, in *Advances in Chromatography*, Vol. 1, J. C. Giddings and R. A. Keller (eds.), Dekker, New York, 1965, p. 113.

26. E. Stahl and H. K. Mangold, in *Chromatography*, 2nd ed., E. Heftmann (ed.), Reinhold, New York, 1967, Chapter 8.

27. R. Maier and H. K. Mangold, in *Advances in Analytical Chemistry and Instrumentation*, Vol. 3, C. N. Reilley (ed.), Wiley–Interscience, New York, 1964, p. 369.

28. J. M. Bobbitt, *Thin-Layer Chromatography*, Reinhold, New York, 1963.

29. J. C. Touchstone, S. S. Levin, and T. Murawec, *Anal. Chem.* **43**, 858 (1971).

30. P. Kabasakalian and J. M. Talmage, *Anal. Chem.* **34**, 273 (1962).

31. E. Soczewinski, in *Advances in Chromatography*, Vol. 8, J. C. Giddings and R. A. Keller (eds.), Dekker, New York, 1969, p. 91.

32. R. E. Majors, *Anal. Chem.* **45**, 755 (1973).

33. F. Eisenbeiss, *Chem. Zeit.* **95**, 237 (1971).

34. K. J. Bombaugh, in *Modern Practice of Liquid Chromatography*, J. J. Kirkland (ed.), Wiley–Interscience, New York, 1971, p. 350.

SELECTED BIBLIOGRAPHY

Block, R. J., E. L. Durrum, and G. Zweig, *A Manual of Paper Chromatography and Paper Electrophoresis*, 2nd ed., Academic Press, New York, 1958.

Bobbitt, J. M., *Thin-Layer Chromatography*, Reinhold, New York, 1963.

Kirchner, J. G., *Thin-Layer Chromatography* (*Technique of Organic Chemistry*, Vol. 12), Wiley–Interscience, New York, 1967.

Lederer, E., and M. Lederer, *Chromatography*, Elsevier, New York, 1957 (PC and column).

Randerath, K., *Thin-Layer Chromatography*, 2nd ed., Academic Press, New York, 1963.

Sherma, J., and G. Zweig, *Paper Chromatography*, Academic Press, New York, 1971.

Stahl, E., *Thin-Layer Chromatography*, Springer–Verlag, New York, 1965.

Truter, E. V., *Thin-Layer Chromatography*, Clever–Hume Press, London, 1962.

ZONE ELECTROPHORESIS 16

Electrophoresis is carefully defined in Chapter 2 because of the confusing terminology that is used. It is also known as zone ionophoresis, electropherography, ionography, electromigration, electrochromatography, and electrochromatophoresis (1). Although zone electrophoresis has some operational similarities with chromatography, it is not a chromatographic technique. Other names that are used are intended to describe a particular system, such as gel electrophoresis, disc electrophoresis, and immunoelectrophoresis. However, this chapter describes only the fundamentals of electrophoresis in its most common form, zonal electrophoresis on a support. The term "zonal" has been defined as a batchwise sampling technique, even though most chemists associate the term with the use of a support when they refer to electrophoresis.

The original work in electrophoresis by Tiselius (2) was unsupported, or "free," using solutions whose boundaries were caused to move under the influence of a voltage gradient, it thus became known as the "moving boundary" method. For his invention in 1937 Tiselius was awarded a Nobel prize in 1948.

The concept of ion mobility, θ, was introduced in Chapter 4; two fundamental equations were presented:

$$\theta = \frac{q^{\pm}}{6\pi r \eta} \tag{1}$$

and

$$s = \frac{\theta E t}{L} \tag{2}$$

where s is the distance migrated by a solute of charge $q^{\pm}$ and radius, r; E is the voltage applied across a support of length L; η is the viscosity of the liquid phase; and t is the time of development. The theory of electrophoresis is not well developed and can be used for qualitative or semiqualitative purposes only.

THEORY

Ion Mobilities

Equations 1 and 2 permit us to draw some general conclusions about electrophoresis. Most obvious is that the ionic mobility is directly proportional to the ionic charge, $q^{\pm}$. Thus the easiest separation is a cation from an anion since they will migrate in opposite directions. Furthermore, a divalent ion will migrate twice as far as a univalent ion, all else being equal. In general, the distance of separation, Δs, of two ions A and B of the same charge is

$$\Delta s = s_A - s_B = (\theta_A - \theta_B)\frac{tE}{L} \tag{3}$$

To achieve a good separation (large Δs), large differences in mobility are desirable as well as a long time t and a large voltage gradient (E/L).

Equation 2 indicates that radius r and viscosity η are two other parameters that affect mobility and hence separability. The viscosity of the solvent is the same for both ions in a given system, but low-viscosity solvents are preferred to increase mobility and decrease time. The radii of the two ions should be as dissimilar as possible. Actually, equation 2 assumes spherical ions, but the shape is also a variable. Furthermore, ion mobility is restricted by the tortuous path an ion must follow in a support. Additional discussion about these and other effects can be found in the review by Edward (3).

Other Factors

Some other factors that are important in achieving separations do not appear in equations 1 and 2. The one that is used to greatest advantage is the degree of ionization.

Degree of Ionization. The solvent used in electrophoresis is often a buffer selected to take advantage of the equilibrium exhibited by weak acids and bases. Take, for example, the well-known ionization of acetic acid (HOAc):

$$HOAc + H_2O \rightleftharpoons H_3O^+ + OAc^- \tag{4}$$

The pK for this reaction is 4.76; so at a pH of 4.76, 50% of the acid will be present as unionized HOAc and 50% as the anion OAc$^-$. At lower pH values, less of the acid is ionized. At a pH of about 2, it can be assumed to be entirely in the molecular forms, so it will not move in a voltage gradient. On the other

Table 1. Ionization Constants and Electrophoretic Behavior of Histidine and Glutamic Acid

| Amino Acid | Ionization Constants | | | | Relative Electrophoretic Migration at pH[a] | | |
	pK_1 (Protonated)	pK_2	pK_3	Isoelectric Point	3.3	7.2	9.3
Histidine	6.10	9.18	—	7.7	−68	−8	+10
Glutamic acid	2.30	4.28	9.67	3.3	+7	+72	+77

[a] Data from reference *1*.

hand, as the pH is increased above 4.76, the percentage of ionization increases up to 100% at about pH 8. Thus at pH 2 the mobility of acetic acid is zero, and at pH 4.76 it is about one-half its mobility at pH 8.

To separate two weak acids by electrophoresis, one chooses the pH at which the greatest difference in ionization and mobility occurs. This has been found to occur at the value given in the following equation:

$$\text{pH} = \frac{pK_A + pK_B}{2} - \log\left[\frac{(\theta_A/\theta_B)^{\frac{1}{2}} - (K_A/K_B)^{\frac{1}{2}}}{1 - (\theta_A K_A/\theta_B K_B)^{\frac{1}{2}}}\right] \tag{5}$$

Another interesting example is the amino acids, which have both acidic and basic properties (amphoterism). Consider as an example histidine and glutamic acid, whose pK values are given in Table 1. Obviously glutamic acid is the stronger acid: it has two carboxylic acid groups, whereas histidine has only one. The pK_1 values are for the protonated form; that is, at a pH equal to pK_1, the amino acid is half in the molecular form and half in a cationic form. Similarly, at a pH equal to pK_2, a given amino acid is half in the molecular form and half in the anionic form. At some point between pK_1 and pK_2, the amino acid must be 100% molecular; this is known as the isoelectric point. At their isoelectric points amino acids exist as "zwitterion," which have no net charge and will not migrate under voltage. A representative zwitterion formula is

The electrophoretic data in Table 1 reflect these principles for the two amino acids. (Note that the migrations are *relative* values at each pH.) At the first pH, 3.3, the glutamic acid is very close to its isoelectric point and migrates very little; histidine is virtually completely protonated and cationic, so it moves very far toward the negative electrode. At pH 7.2, histidine is near its isoelectric point and moves very little; glutamic acid, having lost its carboxylic proton, is anionic and moves toward the positive electrode. Finally, at pH 9.3 both amino acids are negatively ionized and move toward the positive electrode. Note, however, that glutamic acid, some of which is in the divalent form, moves farther than monoionic histidine.

In this example, any of the three pH values would be satisfactory for separating these two amino acids. In actual practice, with a large number of compounds to be separated, the choice would be more restricted.

Complex Formation. By forming complexes between a solute and a component of the solvent (buffer) it is possible to form ions from molecules and to change the sign of an ion from positive to negative. The major example of the first type is the carbohydrates, which do not ionize readily. However, they do form ionic complexes with the borate ion in basic solution, so they can be separated by electrophoresis. Anionic complexes can be formed with several other inorganic oxyacids (*1*).

Some complexes can be formed that will result in a change in the sign of the charge. Consider, for example, the complex formed between a metal ion and a ligand like the chloride ion. At a low chloride-ion concentration the metal ion will be cationic; as the concentration of chloride is increased, a point will be reached where the anionic chloro complex will predominate. The sign of the charge will depend on the ligand concentration and the stability constant, so a separation of metal ions can be based primarily on their differences in stability constants. This use of a complex-formation reaction is very similar to the procedure used in ion-exchange chromatography.

Electrolyte Concentration. So far we have assumed the presence of an electrolyte in the solvent. It has served as a buffer and/or as a complexing agent. In general, an electrolyte is necessary to carry the current and lower the resistance of the support.

However, a high concentration of electrolyte may be disadvantageous because ionic mobility is inversely proportional to the square root of the ionic strength. As ionic strength increases, the mobility decreases, which is undesirable. The reason for the decreased mobility is that an ion experiences a smaller local voltage gradient at a high ionic strength due to the aggregation around it of counterions, which shield it. For example, a layer of anions will surround a given cation, and a more diffuse layer of cations will probably surround the anionic layer. These layers are different from, and should not be confused with, the actual formation of anionic complexes as discussed in the last section.

For particles of colloidal size (such as proteins) this effect is known as the Helmholtz double layer. Because the layers are of opposite charge, a potential, called the zeta potential, can be assumed to exist between them. This potential affects the electrophoretic mobility of colloids as given in the following equation (4):

$$\theta = \frac{\zeta \iota}{4\pi\eta} \qquad (6)$$

where ζ is the zeta potential and ι is the dielectric constant. The ionic mobility increases as the zeta potential increases, but a high electrolyte concentration tends to negate the double layer and decrease the zeta potential and the ionic potential.

Electroosmosis. The concept of the Helmholtz double layer is useful in explaining another effect in electrophoresis—electroosmosis. The double layer arises when a porous solid is in contact with a liquid; the surface of the solid acquires a negative charge and attracts a double layer of counterions from the solution around it. When a potential is applied across such a system, a flow of solvent results. This model describes the electrophoresis system and explains the observed flow of un-ionized solvent. It is called electroosmosis.

When the solvent moves, it carries solute ions and molecules with it. This flow is not intended to occur in electrophoresis and complicates the system, so some correction must be made. Usually this is done by observing the movement (if any) of an uncharged molecule and correcting the movement of the ions by an equal amount (added to the anions and subtracted from the cations).

Diffusion. Working against separation is longitudinal molecular diffusion, which will occur during the time the solute is in the solvent (on the support). In Chapter 4 we saw that band broadening (expressed as the quarter-band width σ) is directly proportional to the square root of the time of diffusion. This zone broadening is another reason for keeping the analysis time short.

Adsorption. Strictly speaking, adsorption is not a part of the electrophoretic process, but in practice some solutes do adsorb on the support. When this occurs, it will decrease the expected ion mobility and usually will result in tailing bands. It is to be avoided if at all possible.

APPARATUS AND TECHNIQUES

Experimentally electrophoresis is somewhat similar to plane chromatography. The electrophoresis bed or support can be paper or a thin layer as in plane chromatography, and it is usually enclosed in a chamber. Sample application and quantitation techniques are likewise similar. Consequently both techniques are included in a "teacher's guide" of apparatus and experiments published in 1965 (5). Other information on apparatus can be obtained from the monographs listed in the bibliography.

Chamber and Support

Many different types of medium have been used for the support or bed, but the most popular one is paper. Others are thin-layer plates (using all types of solid), cellulose acetate, starch grains, starch gel, and acrylamide gel. The geometrical shape can vary from a plane or a strip to a cylinder, but the former is most common.

Whitaker (1) lists four ways a bed can be arranged. Take, for example, a paper strip. It can be (1) unsupported or suspended over a rod at its center; (2) supported on a glass or plastic plate and enclosed in a large chamber; (3) supported and enclosed on both sides by glass or plastic plates; or (4) immersed in an inert liquid. Methods 2 and 3 are most common (see, for example, Chapter 2 in reference 1).

Other Experimental Parameters

The electrolyte is often a buffer and should have a concentration of about 0.1 M. A large volume is desirable so that electrode reaction products will be less likely to diffuse onto the bed. This is also prevented by installing a barrier between the section of the electrolyte containing the electrode and the section containing the end of the bed or strip. Two electrolyte reservoirs are needed, one for each electrode, and their volumes must be exactly equal to prevent a siphon effect on the bed. The electrodes can be carbon rods, but platinum is preferred.

Voltage. The voltage commonly used is 400 V, which produces a gradient of 10 V/cm on a 40-cm strip. A typical current is 100 mA. Such a system uses 40 W and produces considerable heat. A cooling mechanism is therefore desirable to prevent excessive temperatures and solvent evaporation. The latter causes an increase in buffer concentration, which allows a larger current flow and thus produces more heat. As a minimum the system should be covered and presaturated. Then, if the analysis time is not too long, the temperature increase can be tolerated.

Higher voltages of about 4000 V have been shown to be advantageous (6). Since a high voltage will speed up the mobility and hence the analysis time, some separations are improved because there is less time for diffusion broadening. This is especially true for small solute molecules, which have large coefficients, as shown in Figure 1. The amino acid separation is much better in 30 minutes (at 4000 V) than at 5 hours (at 400 V), but the large proteins show little broadening even in 10 hours. Of course, the total analysis

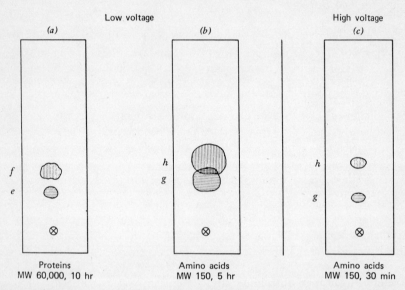

Figure 1. Example of low-voltage (400 V—10 V/cm) versus high-voltage (4000 V— 100 V/cm) electrophoresis separation on 40-cm paper. From R. K. Vitek, *Am. Lab.* **2** (1970). Copyright by International Scientific Communications, Inc.

time is a factor itself, and the decreased time for the high-voltage procedure is also an advantage. Larger amounts of heat are generated, however, and cooling is necessary, and the higher voltage is a safety hazard.

Sampling.   Samples can be applied with a micropipette or microsyringe as in plane chromatography. Two methods are possible. First, the strip can be wetted with the solvent and placed in a saturated chamber before the sample is applied, but as a consequence the sample spot will be wider when applied to a wet strip. Alternatively, the sample can be spotted on the dry strip and the spot kept small. In this case, however, the strip cannot be pre-equilibrated. Also, it should be noted that the strip containing the sample should be immersed in solvent and wetted all at once rather than allowed to become wet slowly by capillary action from the ends. The latter procedure may result in some sample elution if the solvent from the two ends does not meet exactly in the middle of the strip.

Techniques

Normal development is in one dimension, with the sample starting either at the center or at one end (if migration is in one direction). Two-dimensional techniques have been used the same as in plane chromatography. One dimension is usually by electrophoresis and one by chromatography, the combination being highly effective.

Qualitative and quantitative analysis techniques are very similar to those described for plane chromatography.

SUMMARY AND EVALUATION

Zone electrophoresis has been applied to inorganic ions and to a large number of organic and biochemical compounds, from amino acids to vitamins. Most of the separation conditions have been arrived at empirically, which indicates that many of them are easy to perform. On the other hand, some of the specialized techniques, such as the use of starch gels, require considerable skill and practice. The major advantages and disadvantages are listed in Table 2.

Table 2. Evaluation of Zone Electrophoresis

Advantages	Disadvantages
1. Wide applicability for samples that can be ionized, including large molecules	1. High voltage hazardous
2. Can be combined with plane chromatography for greater selectivity and efficiency	2. Not too efficient
3. Faster than some equivalent separations, for example, gel permeation	3. Difficult to quantitate
4. Apparatus and technique often simple	

REFERENCES

1. J. R. Whitaker, *Electrophoresis in Stabilizing Media*, Vol. 1 of a two-volume series, Academic Press, New York, 1967.
2. A. Tiselius, *Trans. Faraday Soc.* **33**, 524 (1937).

3. J. T. Edward, in *Advances in Chromatography*, Vol. 2, J. C. Giddings and R. A. Keller (eds.), Dekker, New York, 1966, p. 63.

4. R. J. Akers, *Am. Lab.* **4**[6], 41 (1972).

5. I. Smith, in *Advances in Chromatography*, Vol. 1, J. C. Giddings and R. A. Keller (eds.), Dekker, New York, 1965, p. 61.

6. R. K. Vitek, *Am. Lab.* **2**[10], 62 (1970).

SELECTED BIBLIOGRAPHY

Bier, M. (ed.), *Electrophoresis*, Academic Press, New York, Vol. I, 1959, Vol. II, 1967.

Block, R. J., E. L. Durrum, and G. Zweig, *A Manual of Paper Chromatography and Eectrophorlesis*, 2nd ed., Academic Press, New York, 1958.

Heftmann, E., (ed.), *Chromatography*, 2nd ed., Reinhold, New York, 1967, Chapters 10 by R. J. Wieme and Chapter 11 by H. Michl.

Karger, B. L., L. R. Snyder, and C. Horvath, *An Introduction to Separation Science*, Wiley–Interscience, New York, 1973, Chapter 17 by M. Bier.

Whitaker, J. R., *Electrophoresis in Stabilizing Media*, Academic Press, New York, 1967.

Wieland, T., and K. Dose, in *Physical Methods in Chemical Analysis*, Vol. III, W. G. Berl (ed.), Academic Press, New York, 1956, p. 29.

DIFFERENTIAL DIALYSIS 17

Dialysis is seldom used in the analytical laboratory to separate a multi-component sample. Its main use is in concentrating, desalting, and purifying such materials as proteins, hormones, and enzymes. It is a simple technique and one usually performed with fairly large quantities (preparative scale). A typical apparatus is shown in Figure 1; the solution to be purified (the retentate or dialyzate) is put inside the membrane bag, which is immersed in the diffusate; the latter is either stirred or circulated or both. This method is slow and not very selective; it is not the technique we want to consider further.

Some attempts have been made, however, to increase the speed of dialysis as well as its selectivity. These efforts are worthy of brief mention, and they will be called differential dialysis methods. Most of this chapter is devoted to the work of Craig and co-workers, who have produced an apparatus for thin-film countercurrent dialysis that has many of the same operational features as the techniques of countercurrent liquid–liquid extraction and chromatography (1). This similarity to other separation methods is the justification for including differential dialysis in this text.

Another new development deserving mention is a new type of membrane in the form of hollow fibers (2). For classical dialysis and ultrafiltration, they offer the advantages of speed and compact size.

THEORY

Differential dialysis as used in analytical separations is the combination of two mechanisms. First is the *kinetically controlled diffusion* discussed in Chapter 4. When a concentration gradient $(C_o - C_i)$, exists on opposite sides of a semi-permeable membrane, the amount of solute, N, that can diffuse through it is

$$N = kA(C_o - C_i)t \qquad (1)$$

where k is the permeability coefficient, A is the area of the membrane, and t is the time.

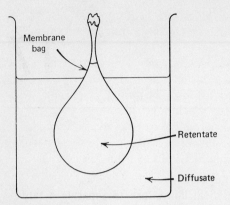

Figure 1. Classical dialysis apparatus.

The other effect is *mechanical sieving*, which occurs when some of the pores of the membrane are smaller than some of the solute molecules in the sample. Consequently selective diffusion takes place because the larger molecule cannot pass through the membrane even though a concentration gradient exists. The phenomenon is somewhat analogous to gel permeation except that in dialysis the small molecules go through the membrane, whereas in GPC they diffuse in and out of the pores of the stationary phase. Otherwise the two mechanisms are quite similar.

Typical data illustrating selective dialysis are shown in Figure 2 for a separation of glucose and sucrose (*3*). The figure shows that the diffusion process was a first-order one for both solutes and that a fairly good separation could be achieved with this particular system. Note, however, the long times required.

If the sample contains electrolytes, some of which are not able to pass through the membrane, the Donnan membrane theory (Chapter 4) predicts an unequal distribution of the diffusable electrolyte on the two sides of the membrane.

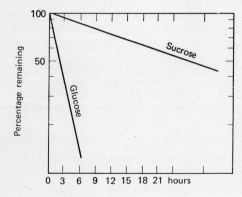

Figure 2. Example of selective dialysis. Reprinted from L. C. Craig, *Science* **144,** 1093 (1964). Copyright 1964 by the American Association for the Advancement of Science.

Three other factors that complicate this simple mechanism are adsorption, the presence of an electrokinetic charge on the membrane, and osmosis (4). The membrane itself is a major factor in determining the extent of all three of them, and the solute affects the degree of adsorption and osmosis as well.

Although osmosis complicates the diffusional processes, it forms the basis of a related technique that is very useful: reverse osmosis, or ultrafiltration. In this process pressure is applied to a solution, thus forcing the *solvent* molecules to permeate a membrane separating it from another chamber. The membranes and other apparatus required are similar to those used in dialysis. For further discussion and laboratory procedures, see reference 5.

APPARATUS AND PROCEDURES

Craig's thin-film countercurrent dialyzer (1) is shown in Figure 3. The inner glass tube, which has a diameter of about 17 mm, has a wetted, seamless casing or membrane stretched over it. The retentate (the phase into which the sample is placed) is pumped down the center of the glass tube, and it passes up the outside of the tube between it and the membrane.

This assembly is placed in an outer glass tube, which has an inside diameter of about 18 mm. The diffusate enters at the top and passes over the membrane, exiting at the bottom. This phase can be moving (true countercurrent operation) or stationary (pseudocountercurrent). Both phases can be collected, if desired. The outer tube is rotated to improve mixing.

Obviously this arrangement has the desirable characteristics of a large membrane area and a thin retentate film. Furthermore, the membrane is stretched very tightly, so that it is thin and provides fast diffusion rates. A membrane for this apparatus must be very strong, and cellulose has been used.

Other membrane materials are cellophane, collodion, parchment, gelatin, and synthetic polymers. Cellophane is very good because it has practically no fixed charge and shows minimal adsorption. The sizes of the pores in the membrane can be varied by selective stretching, acetylation, and swelling with zinc chloride (6). Acetylation also strengthens cellulose.

In operation, the retentate is pumped at 0.5 to 0.8 ml/min; the diffusate can be stationary or move at up to 10 ml/min under the influence of gravity. The sample is introduced into the retentate stream, and both streams are monitored. This procedure would seem to be optimal for differential dialysis, but it is still not efficient enough to be of much use for multicomponent separations. Gel-permeation chromatography does a better job of separating molecules on the basis of size (molecular weight).

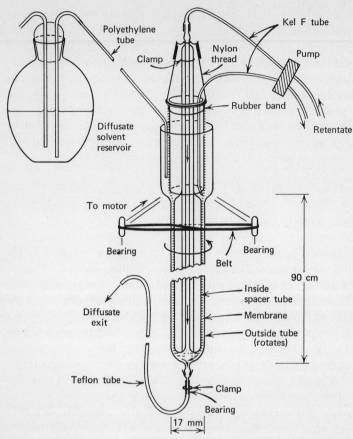

Figure 3. Schematic drawing of thin-film countercurrent dialyzer. Reprinted with permission from L. C. Craig and H. Chen, *Anal. Chem.* **41,** 590 (1969). Copyright by the American Chemical Society.

SUMMARY

It is interesting to compare separation techniques involving the countercurrent flow of two liquids. They are liquid–liquid extraction (LLE), liquid–liquid chromatography (LLC; both in columns and on plane surfaces), and liquid–liquid dialysis, plus the new technique of countercurrent chromatography (Chapter 11). In every case except dialysis the two liquids must be immiscible; in dialysis a membrane keeps them separate. In chromatography an inert support is used to keep one phase stationary, and in the other techniques this is accomplished by the design of the equipment, aided by the membrane in the case of dialysis. In practice these different techniques find different uses, but they are best compared and studied on a unified basis.

REFERENCES

1. L. C. Craig and H. Chen, *Anal. Chem.* **41,** 590 (1969).
2. W. F. Blatt, *Am. Lab.* **4**[10], 78 (1972).
3. L. C. Craig, *Science* **144,** 1093 (1964).
4. C. W. Carr, in *Physical Methods in Chemical Analysis*, Vol. 4, W. G. Berl (ed.), Academic Press, New York, 1961, p. 7.
5. G. R. Garbarini, R. F. Eaton, T. K. Kwei, and A. V. Tobolsky, *J. Chem. Educ.* **48,** 226 (1971).
6. L. C. Craig, in *Advances in Analytical Chemistry and Instrumentation*, Vol. 4, C. N. Reilley (ed.), Interscience, New York, 1965, p. 35.

SELECTION OF A SEPARATION METHOD

18

A major objective of this book has been the presentation of material in a unified fashion to facilitate the comparison of the major methods of separation. If this approach has been successful, its culmination should be the ability to select a good separation method for a given analytical problem. This chapter attempts to summarize the most important criteria for the selection of a method.

There is, of course, no substitute for experience, and this brief chapter will not provide that. The wisdom of experienced experts is available however, in the literature. Its use should not be overlooked or underestimated.

TEN CRITERIA

The 10 criteria discussed in this section are among the most important ones in the choice of a separation method. The order in which they are presented is, in most instances, an order of decreasing importance, although one of the criteria (such as cost) could be of overriding importance. The first four criteria are characteristics of the sample itself, and the last six are requirements of the analysis.

The first eight criteria are listed in Table 1, which serves to classify the major separation methods. An "A" or "B" in the table indicates that this method is limited to this particular purpose (or sample) or it is most commonly used for this purpose (or this type of sample). An "X" indicates that both purposes (sample types) are applicable or that the distinction is not important for this method. Not all classifications are as clear-cut as the table would indicate, and some of them were chosen on the basis of the most common usage rather than the potential usage. For example, LSC can certainly be used for preparative separations, but the type of high-pressure LSC described in Chapter 14 is used primarily for analytical separations at present. Nevertheless, a glance at the table will reveal that no two methods have exactly the same classification, although several are quite close.

Table 1. Classification of the Major Separation Methods According to 10 Criteria

Criteria		Separation Method[a]											
A	B	LLE	D	GC	LSC	LLC	IEC	GPC	PC	E	DL	P	IC	M
1. Hydrophilic	Hydrophobic	X	B	B	B	B	A	X	X	A	A	A	X	A
2. Ionic	Nonionic	X	B	B	B	B	A	B	X	A	X	A	B	X
3. Volatile	Nonvolatile	B	A	A	B	B	B	B	B	B	B	B	X	B
4. Simple	Complex	A	A	B	B	B	B	B	A	A	A	A	A	A
5. Quantitative	Qualitative	A	B	A	A	A	A	A	B	B	B	A	A	A
6. Individual	Group (type)	X	A	A	X	A	A	X	B	X	B	B	X	X
7. Recovery	Purity	X	A	B	B	B	B	B	B	B	A	A	B	B
8. Analytical	Preparative	X	B	A	A	A	A	A	A	A	B	X	B	X

[a] Abbreviations: LLE, liquid–liquid extraction; D, distillation; GC, gas chromatography; LSC, liquid–solid chromatography; LLC, liquid–liquid chromatography; IEC, ion-exchange chromatography; GPC, gel-permeation chromatography; PC, plane chromatography; E, electrophoresis; DL, dialysis; P, precipitation; IC, inclusion compound; M, masking.

Characteristics of the Sample

The first two characteristics are often related; they are (1) the hydrophilic or hydrophobic nature and (2) the ionic or nonionic character of the sample. Most techniques can handle either the ionic hydrophilic type of sample or the nonionic hydrophobic type, but not both. The notable exceptions in Table 1 are LLE and plane chromatography.

The third sample characteristic is its volatility. Distillation and GC are restricted, for the most part, to volatile samples, and they will probably be the preferred separation methods for such samples. It is possible, of course, to use other methods for volatile samples. Related to the sample's volatility are its thermal stability and overall reactivity. Distillation and GC are restricted to compounds that are stable at the elevated operating temperatures. General chemical reactivity is more difficult to generalize, and for each sample, consideration should be given to the probability of chemical reaction with the components of the system. Solid surfaces such as those in GSC, LSC, plane chromatography, and zone electrophoresis are likely to be active or catalytic.

The final sample characteristic is its complexity in terms of the number of components. The two extremes are listed in the table as simple and complex. Only the chromatographic techniques are efficient enough to be capable of separating complex mixtures. Within a given classification (A or B) a further consideration might be specific interferences or matrix effects.

Requirements of the Analysis

The first of the requirements of analysis is the relationship of the separation step to the identification or measurement step which follows in a qualitative or a quantitative analysis, respectively. We have already noted that the separation method best suited to quantitation at present is GC. Also popular, but for a totally different type of sample, is precipitation, the separation step most commonly used in gravimetry. In a quantitative analysis the requirements of the analysis may also influence the choice of the separation step. Some examples are the accuracy and precision desired and the amount of the desired constituent in the sample.

The second requirement is the degree of separation required. Some analyses are intended to separate all of the individual components from each other, while for other purposes this may be more specificity than is desirable. For example, in one analysis it may be desirable to know the number (and concentration) of each hydrocarbon present, but in another analysis it may be sufficient (even preferable) to how much of each type—paraffin, olefin, aromatic—is present. This is a difficult classification to make because many methods can be adapted to both requirements. The table indicates the most common of the two possibilities for each method, but many more could be indicated by "X."

Theoretically it is impossible to achieve both a good recovery and a high purity, so this choice is listed next. In Chapter 10 some data were given to show that the crosscurrent technique is preferable for high recovery and the countercurrent technique is preferable for high purity. Chromatography and electrophoresis are countercurrent techniques, and they are listed in the table as methods giving high purity. Differential dialysis is described in Chapter 17 as a countercurrent technique, but one that is not very efficient. Therefore in Table 1 its classification is based on the more common, classical usages such as desalting, which favor recovery.

The final requirement listed in the table refers to the quantity of sample one wants to separate—a small, analytical sample or a large, preparative one. As already noted, the most common usages are indicated in the table and some flexibility is possible. By and large, however, the chromatographic techniques are troublesome to scale up to handle large quantities. A related limitation regarding the quantity of material to be separated is the amount of sample available. For example, if only 10 μl of material is available, a distillation is out of the question.

The two criteria not included in the table are (1) cost and speed, and (2) personal preference and availability of equipment. Cost refers to the cost per analysis in terms of the time required and thus is related to the number of samples to be separated. However, cost is also a factor in the other criterion relative to the purchase of equipment. Personal preference is related to previous experience and is a very important consideration in many choices.

APPROACHING A PROBLEM

An obvious first step is to note the state of the sample. Gases can only be separated by GC and the formation of inclusion compounds (of those we have considered). Other possibilities are thermal diffusion and chemical reactions (and "irreversible" adsorption). If distillation is to be used, the sample must be a liquid. For solids, the equivalent choice is zone refining. Both liquids and solids can be separated by all of the other methods including GC, subject to the volatility requirement (No. 3 in Table 1).

The next step might be a quick screening of the sample if its composition is unknown. The best screening techniques are distillation (for liquids) and TLC (for liquids or solids). If the entire sample is volatile, programmed-temperature GC can also be used for screening. If the sample is aqueous, special stationary phases (such as Porapak) should be used in GC and electrophoresis can be used for screening.

For a more complete separation, a consideration of the first four criteria will usually indicate the best one or two methods. If the choice is between several chromatographic methods, the selection of the best one might be based on the molecular weight of the sample. Gas chromatography is best for low molecular weights (to several hundred), LSC and LLC are best for medium molecular weights (several hundred to several thousand), and GPC is best for high molecular weights.

One final suggestion for complex samples is to use more than one separation method. We have already noted the advantages of two-dimensional plane chromatography and the combination of zone electrophoresis and plane chromatography. A simple batch extraction often serves as the first step in a multiple-method separation. Other combinations could be mentioned, but the principle should be clear.

APPENDIX

TYPICAL CHROMATOGRAPHIC CALCULATIONS

Figure 1 shows a liquid chromatographic separation of acetylacetonate chelates of beryllium, chromium, ruthenium, and cobalt. The conditions of the separation were:

Stationary phase: a ternary mixture of 64% ethanol, 34% water, and 1.6% isooctane.

Mobile phase: a ternary mixture of 98% isooctane, 2% ethanol, and 0.08% water

$$V_M = 1.32 \text{ ml.}$$
$$V_S = 0.142 \text{ ml.}$$
$$F = 0.291 \text{ ml/min.}$$
$$\text{Column length, } L = 23 \text{ cm.}$$
$$\text{Inside diameter of column} = 2.7 \text{ mm.}$$
$$\text{Inlet pressure, } P_i = 50 \text{ atm.}$$
$$\text{Chart speed} = 1 \text{ cm/min.}$$

Calculate β for this column, and for both the beryllium and the cobalt chelates calculate the following parameters: V_R, V_N, k', R_R, K_p, n, and H. For the unresolved mixture of chromium and ruthenium chelates, calculate the resolution, R_s, assuming that the peaks are of equal height.

The partition coefficients for these chelates have been determined by static methods and found to be as follows: beryllium, 4.4; chromium, 16.1; ruthenium, 23.2; and cobalt, 33.5.

Figure 2 shows a gas chromatogram of toluene. The operating conditions and some other necessary data are given below:

Column: 48 × 0.25 in. (outside diameter); inside diameter = 5.0 mm; packed with 12.0 g containing 20% (w/w) dinonyl phthalate on 80/100 mesh Chromosorb P

$$\text{Column temperature} = 100°C$$
$$\text{Ambient temperature} = 25°C$$
$$\text{Ambient pressure } P_0 = 760 \text{ torr}$$

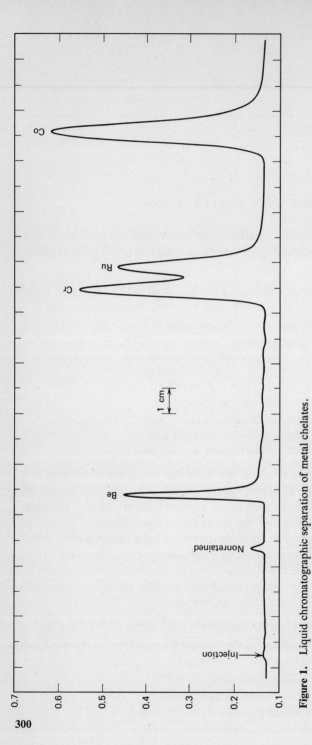

Figure 1. Liquid chromatographic separation of metal chelates.

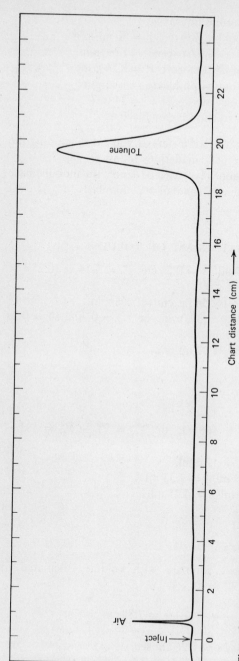

Figure 2. Gas chromatogram of toluene.

Outlet flow measured at ambient conditions
with a soap-bubble flowmeter $= 80$ ml/min
Chart speed $= 1$ in./min
Density of Chromosorb P $= 2.26$ g/ml
Density of dinonyl phthalate $= 1.03$ g/ml
Vapor pressure of water at 25°C $= 24$ torr
Calculated pressure correction factor $j = 0.65$

Calculate as many chromatographic parameters as you can, including the total volume of the column. The actual calculations follow.

Note that in this example the symbol V_G is used to denote the mobile phase (which is a gas) rather than the general symbol V_M. Similarly, V_L is used rather than V_S for the stationary phase.

CALCULATIONS: GAS CHROMATOGRAM OF TOLUENE

1. Total volume of column, V_t. Radius $= 0.25$ cm. $V_t = 3.14 \times (0.25)^2 \times 4 \times 12 \times 2.54 = 23.9$ ml.
2. $t_R = 19.45$ cm $\times 1$ min/in. $\times 1/2.54$ in./cm $= 7.66$ min.
3. $t'_R = (19.45 - 0.70) \times 1/2.54 = 7.39$ min, or $t_M = 0.70/2.54 = 0.28$ and $t'_R = 7.66 - 0.28 = 7.38$ min.
4. $F_c = 80 \times 373/298 \times 736/760 = 97$ ml/min.
   $\bar{F}_c = 97 \times 0.65 = 63$ ml/min.
5. $V_R = 7.66 \times 97 = 743$ ml.
6. $V'_R = 7.39 \times 97 = 716$ ml.
7. $V_R^0 = 743 \times 0.65 = 482$ ml.
8. $V_N = j \times V'_R = 716 \times 0.65 = 466$ ml, or $V_N = V_R^0 - V_G^0 = 482 - 18 = 464$ ml.
9. $V_G^0 = t_M \times \bar{F}_c = 0.28 \times 63 = 17.6$ ml.
10. $V_L = 12$ g $\times 0.20 \times 1/1.03$ ml/g $= 2.33$ ml.
11. $V_{SS} = 12 \times 0.80 \times 1/2.26$ ml/g $= 4.25$ ml.

$$V_G^0 = 17.6 \qquad 72.8\% = \varepsilon_T$$
$$V_L = 2.3 \qquad 9.5\%$$
$$V_{SS} = \underline{4.3} \qquad 17.7\%$$
$$24.2 = \text{total volume; compare with volume calculated in item 1.}$$

12. $\bar{u} = \dfrac{\bar{F}_c}{A \times \varepsilon_T \times 60} = \dfrac{63}{0.182 \times 0.73 \times 60} = 7.88$ cm/sec, or

   $\bar{u} \approx L/60\, t_M = 122/60 \times 0.28 = 7.26$ cm/sec.
13. $\beta = 17.6/2.33 = 7.6$.
14. $K_p = V_N/V_L = 465/2.33 = 199$.

15. $k' = K_p/\beta = 199/7.6 = 26.2$.

Also note: $V_R^0 = V_G^0(1 + k') = 17.6\,(27.2) = 479$ ml (Compare with item 7).

16. $R_R = V_G^0/V_R^0 = 17.6/482 = 0.0366$, or toluene spends 3.7% of its time in the mobile phase.

$(1 - R_R) = 0.9634$, or toluene spends 96.34% of its time in the stationary phase.

$$k' = \frac{1 - R_R}{R_R} = 96.3/3.6 = 26.3 \text{ (compare with item 15)}.$$

17. $n = 16(19.45/1.96)^2 = 16(9.93)^2 = 16(99) = 1584$ plates.

18. $H = L/n = 122\text{ cm}/1584 = 0.077\text{ cm} = 0.77\text{ mm}$.

19. $h = H/d_p = 0.77/0.163 = 4.4$ (see Table 5 in Chapter 12 for the average particle diameter).

INDEX

n Number of theoretical plates

n_{eff} = effective number of theoretical plates

n_{lim} = limiting number of theoretical plates

n_t Transfer number

n_{max} = transfer number required to get maximum solute into last funnel

p Partial pressure; for example p_w

p^0 Equilibrium vapor pressure

q Configuration factor (in Rate equation)

q^+ or q^- Electrical charge, plus or minus; also $q^{\pm}$

r Funnel number

r_{max} = funnel number containing solute maximum

s Distance (migration distance, internuclear distance, etc.)

t Time; for example t_R, t_M etc.; see related volume symbols

u Mobile phase velocity (cm/sec)

$\bar{u}$ = average mobile phase velocity

u_F = velocity of mobile solvent front

v Velocity, including solute velocity

$\bar{v}$ = average solute velocity (cm/sec)

w Peak width at the base; for example w_A

w_h Peak width at half of the peak height

x Independent variable or integer

y Dependent variable

z Direction of longitudinal flow

α Separation quotient

β Ratio of phase volumes

γ Activity coefficient

δ Solubility parameter of Hildebrand

ϵ Porosity

ϵ_T Porosity, total

ϵ_I Porosity, interstitial

ϵ^0 Solvent strength parameter of Snyder

ζ Zeta potential

η Viscosity

θ Mobility

ι Dielectric constant

κ Permeability

λ Packing characteristic (rate equation)

μ Dipole moment

ν Reduced velocity

ξ Extent of separation (Rony)

ρ Polarizability

σ Standard deviation or quarter-zone width

τ Surface tension

ϕ Volume fraction of solvent

χ Chemical potential

ψ Tortuosity, obstruction factor (rate equation)

ω Packing factor (rate equation)